MW01503579

Governing the Future

We are living in times of deep and disruptive change. Perhaps the most powerful vector of this change can be described by three related catchphrases: digitalization, artificial intelligence, and dataism. Drawing on considerable expertise from a wide range of scholars and practitioners, this interdisciplinary collection addresses the challenges, impacts, opportunities, and regulation of this civilizational transformation from a variety of angles, including technology, philosophy, cultural studies, international law, sociology and economics. This book will be of special interest to scholars, students, analysts, policy planners, and decision-makers in think tanks, international organizations, and state agencies studying and dealing with the development and governance of disruptive technologies.

Henning Glaser is the Director of the German-Southeast Asian Center of Excellence for Public Policy and Good Governance (CPG), Thammasat University.

Pindar Wong is an Internet pioneer and the former Chairman of VeriFi (Hong Kong) Ltd.

CRC Press Reference Books in Computer Science

CRC Press Reference Books in Computer Science series offers advanced and authoritative resources for students, researchers, and professionals across the field of computer science. Books in this series serve as indispensable references which present rigorously researched and innovative approaches to current and complex issues.

Governing the Future: Digitalization, Artificial Intelligence, Dataism
Henning Glaser and Pindar Wong

Governing the Future
Digitalization, Artificial Intelligence, Dataism

Edited by
Henning Glaser
Pindar Wong

CRC Press
Taylor & Francis Group
Boca Raton London New York

CRC Press is an imprint of the
Taylor & Francis Group, an **informa** business

A CHAPMAN & HALL BOOK

Designed cover image: Shutterstock

First edition published 2025
by CRC Press
2385 NW Executive Center Drive, Suite 320, Boca Raton FL 33431

and by CRC Press
4 Park Square, Milton Park, Abingdon, Oxon, OX14 4RN

CRC Press is an imprint of Taylor & Francis Group, LLC

ISBN: 9781032128382 (hbk)
ISBN: 9781032116372 (pbk)
ISBN: 9781003226406 (ebk)

DOI: 10.1201/9781003226406

Typeset in Minion
by codeMantra

Contents

List of Contributors

Nicholas Agar is Professor of Ethics at the University of Waikato, New Zealand. For the past almost 30 years, he has been exploring the ethical implications of technological change, and the ways in which genetic and cybernetic technologies may alter human beings. He is the author of *Dialogues on Human Enhancement* (Routledge, 2023), *How to Be Human in the Digital Economy* (2019), *The Sceptical Optimist: Why Technology Isn't the Answer to Everything* (2015), and *Truly Human Enhancement: A Philosophical Defense of Limits* (2013). Previously, he worked as professor of ethics and as senior lecturer at the Victoria University of Wellington and was distinguished visiting professor at Carnegie Mellon University, Australia.

Pablo García-Barranquero is a philosopher of medicine, with a special interest in analyzing the process of human aging and the conceptual and normative implications of its (possible) pathologization. He is currently Assistant Professor in the Department of Philosophy at the University of Malaga, Logic and Philosophy of Science. Before joining Malaga, he was post-doctoral researcher in the Department of Philosophy I at the University of Granada, working in a group concerned with the impact of technology on our species. He was also postdoctoral researcher at the Research Unit of Philosophy of Science and Human Development, Universitá Campus Bio-Medico di Roma, working on a project on whether the rejuvenation concept is scientifically and philosophically coherent and plausible. He has also been visiting researcher, among other places, in the School of History, Philosophy, Political Science and International Relations, Victoria University of Wellington, and the Social Medicine Department at UNC.

Russell Buchan is Professor of International Law at the University of Reading and was previously senior lecturer in international law at the University of Sheffield. He specializes in public international law and has particular research and teaching interests in international political and legal theory, UN law, the use of force, international humanitarian law, international criminal law and justice, and international dispute settlement. He is the author of *International Law and the Construction of the Liberal Peace* (2013), *Cyber Espionage and International Law* (2018) and co-author of *Regulating the Use of Force in International Law: Stability and Change* (2021), *Research Handbook on International Law and Cyberspace* (second edition, 2021), and *The Rights to Privacy and Data Protection in Times of Armed Conflict* (2022).

Massimo Durante is Professor of Philosophy of Law and Legal Informatics at the University of Turin. His main interests include law and technology, information ethics, internet governance and information technology law, privacy and data protection law, and artificial intelligence and law. He is the author of *Ethics, Law and the Politics of Information: A Guide to the Philosophy of Luciano Floridi* (2018) and *Computational Power: The Impact of ICT on Law, Society and Knowledge* (Routledge, 2021). He serves as Faculty Fellow of the Nexa Center for Internet and Society at the Polytechnic University of Turin.

Anthony Elliott is Dean of External Engagement at the University of South Australia (UniSA), where he is Bradley Distinguished Professor of Sociology and Executive Director of the Jean Monnet Centre of Excellence in UniSA: Justice and Society. He serves as a member of the Australian Research Council's College of Experts and is a Fellow of the Academy of Social Sciences in the UK, Fellow of the Academy of Social Sciences in Australia, Fellow of the Cambridge Commonwealth Trust, and senior member of King's College, Cambridge. He is the author of *Algorithmic Intimacy: The Digital Revolution in Personal Relationships* (2022), *Contemporary Social Theory: An Introduction* (Routledge, 3rd ed., 2022), and *The Routledge Social Science Handbook of AI* (Routledge, 2021).

Mark Fenwick is Professor of International Business Law at the Faculty of Law, Kyushu University, Japan. His primary research interests are in the fields of white collar and corporate crime, and business regulation in a networked age. Recent publications include *New Technology, Big Data & the Law* (2017, co-edited with Marcelo Corrales and Nikolaus Forgo), *The Shifting Meaning of Legal Certainty in Comparative & Transnational Law* (2017: co-edited with Stefan Wrbka and Matthias Siems), and *Robotics, AI and the Future of Law* (2018: co-edited with Marcelo Corrales and Nikolaus Forgo). He has been Visiting Professor at Chulalongkorn University, Duke University, the University of Hong Kong, Shanghai University of Finance & Economics, the National University of Singapore, Tilburg University, and Vietnam National University.

Robert M. Geraci is Knight Distinguished Chair for Study of Religion and Culture and Professor of Religious Studies at Knox College. Prior to joining Knox College, he was formerly professor of religious studies and faculty director for veteran success at Manhattan College. He is the author of *Apocalyptic AI: Visions of Heaven in Robotics, Artificial Intelligence, and Virtual Reality* (2010), *Virtually Sacred: Myths and Meaning in World of Warcraft and Second Life* (2014), *Temples of Modernity: Nationalism, Hinduism, and Transhumanism in South Indian Science* (2018), and *Futures of Artificial Intelligence: Perspectives from India and the U.S.* (2022). He has been a Visiting Researcher at Carnegie Mellon University's Robotics Institute, the Indian Institute of Science, and the National Institute for Advanced Studies in Bangalore, India. His research has been supported by the US National Science Foundation, the Republic of Korea National Research Foundation, the American Academy of Religion, and two Fulbright-Nehru research awards, and he is a Fellow of the International Society for Science and Religion.

Henning Glaser is a lawyer and mediator, currently serving as the Founding Director of the German-Southeast Asian Center of Excellence for Public Policy and Good Governance (CPG), an independent, interdisciplinary research institute and think tank established by the universities of Münster and Frankfurt/Main (Germany) and Thammasat University (Thailand). He also teaches public law at Thammasat University's Faculty of Law and holds the position of Executive Director and Chairperson of the Board of Directors of the Asian Governance Foundation. His research interests include the future trajectories of the global system, particularly how technology disrupts and determines societal core concepts, identities, and normative structures.

Toni Loh (née Janina Sombetzki; pronouns: none; they/them; neopronouns) is Professor of Applied Ethics – in particular Ethics and Transformation (Angewandte Ethik – insbesondere Ethik und Transformation) at the Center for Ethics and Responsibility and the Department for Social Policy and Social Security at Bonn-Rhein-Sieg University of Applied Sciences. From 2021 to 2024, they worked as an ethicist in a staff position (Stabsstelle Ethik) at the Stiftung Liebenau and, in this role, also managed the Liebenau Foundation's Ethics Committee and the Ethics Cooperation Group (Kooperationskreis Ethik). Among Toni Loh's publications is the first German *Introduction to Trans- and Posthumanism* (2018, fourth edition in 2023) and an *Introduction to Robot Ethics* in German language (2019). Their main research interests lie in the field of trans- and posthumanism (especially critical posthumanism), robot ethics, feminist philosophy of technology, responsibility research, Hannah Arendt, medical and bioethics or ethics in social work and healthcare, theories of judgment, and ethics in the sciences.

Jonathan Pengelly is a freelance data engineer and data scientist specializing in data and technology ethics consulting, with over 15 years of experience working with a diverse range of organizations. Previously, he worked as a data analyst for New Zealand's Ministry of Education and served as a guest lecturer at the University of Wellington, where he earned his PhD in Philosophy.

Elke Schwarz is Reader in Political Theory in the School of Politics and International Relations, Queen Mary University of London. Previously, she was lecturer in international politics at the University of Leicester and held positions as a teaching fellow at University College London (UCL) and lecturer in international relations at Anglia Ruskin University. Her research interests include ethics, digital technology, artificial intelligence, political theory, philosophy of technology, critical security studies, and apocalyptic/post-apocalyptic studies. She is the author of *Death Machines: The Ethics of Violent Technologies* (2018).

Yong Sup Song is Assistant Professor of Christian Ethics and Theology at Youngnam Theological University and Seminary, South Korea. His research interests are focused on the ethical issues of artificial intelligence (AI). He emphasizes the inclusion of regional values and the priority options for the poor in the development of AI. He is working on discovering and introducing cultural values in Korean society for moral AI. He has won a 2022–2023 Collaborative International Research Grant (with Robert M. Geraci, PI) from the American Academy of Religion (AAR). His recent publications include "Marginalization and Transcendence in Transhumanism and Minjung Theology," *Zygon: Journal of Religion and Science* (2023, with Robert M. Geraci); "Evil of Death and Salvation through Faith by Artificial Intelligence," *Main Challenges for Christian Theology Today* (2022, edited by Christophe Chalamet); "Religious AI as an Option to the Risks of Superintelligence: A Protestant Theological Perspective," *Theology and Science* (2021).

Erik P. M. Vermeulen is Professor of Business and Financial Law at Tilburg University, The Netherlands and Senior Legal Counsel at Signify (formerly known as Philips Lighting). His research explores the collision of life, work, and technology. He teaches business organizations and governance, law and technology, and business innovation and finance at Tilburg Law School, TIAS School for Business and Society, and at other universities around the world. He is the co-author of *Organizing-for-Innovation: Corporate Governance in a Digital Age* (2022, together with Mark Fenwick, Toshiyuki Kono, and Tronel Joubert) and co-editor of *Legal Tech and the New Sharing Economy* (2020, together with Marcelo Corrales Compagnucci, Nikolaus Forgó, Toshiyuki Kono, and Shinto Teramoto).

Pindar Wong is an Internet pioneer who served as the first vice-chairman of the Internet Corporation for Assigned Names and Numbers, former chairman of VeriFi (Hong Kong) Ltd, an Internet Financial Infrastructure consultancy, and founder of Hong Kong's Smart Contracts Initiative that pioneered the "Belt and Road Blockchain." He co-founded Hong Kong's first licensed Internet Service Provider in 1993 and was the chairman of the Asia Pacific Internet Association, alternate chairman of Asia Pacific Network Information Centre, and elected trustee of the Internet Society and Commissioner on the Global Commission on Internet Governance.

John Zeleznikow is Professor in both Management and Information Systems at the Institute for Sustainable Industries & Liveable Cities, La Trobe University. He formerly served as Director of The Joseph Bell Centre for Forensic Statistics and Legal Reasoning at the University of Edinburgh Law School, Head of the Department of Computer Science at Latrobe University, and Associate Dean (Research), Faculty of Business and Law at Victoria University. He is the author of nearly 100 refereed journal articles (including Harvard Negotiation Law Review) as well as over 200 refereed conference articles and book chapters. He is the co-author of *Online Family Dispute Resolution: Evidence for Creating the Ideal People and Technology Interface* (2021, together with Elisabeth Wilson-Evered) and of *Enhanced Dispute Resolution through the Use of Information Technology* (2010, together with Arno R. Lodder).

Governing the Technetronic Revolution? An Uncertain Future between Paradise and Pandemonium

Henning Glaser

INTRODUCTION

For several years now, there has been increasing recognition that we are living in an era of profound civilizational change, much of which is unfolding within the realm of science and technology. Although we seem to be acutely aware of the multiple shifts occurring around us, we often grow too easily accustomed to their presence, occasionally forgetting the magnitude of their impact and the consequences they bear. This applies also to the technological revolution, which is unfolding before our eyes. Driven by rapid progress in a whole web of often converging advancements such as digitalization, artificial intelligence (AI), computing, robotics, molecular nanotechnology, bioengineering, and more,[1] contemporary technology has developed an unprecedented capability to impact all facets of human life and fundamentally alter its conditions on a global scale.

From a historical perspective, the present technological revolution possesses several distinctive qualities, setting it apart from all previous technological breakthroughs. This applies to both the process itself and the context in which it is situated.

As a civilizational process, the ongoing technological revolution is unfolding at an unprecedented pace, depth, and breadth. Never have the promises and risks of a technological revolution come with such amplitude while the almost exponential nature of its progress seems to be diminishing our agency over it. The magnitude and impact of the

DOI: 10.1201/9781003226406-1

present technological revolution is reflected by the fact that some even identify it with the coming birth of a new human species, the emergence of *Techno sapiens* and the gradual elimination of *Homo sapiens*.[2]

Regarding the context in which the current technological revolution is situated, its significance can hardly be overstated as well. In fact, the present technological revolution is occurring within an entire world system in flux, at a time when the very foundations of the world seem to be shifting in all important dimensions. This has led to the characterization of the present state of the global system as a "VUCA World," an acronym standing for volatile, uncertain, complex, and ambiguous.[3]

While profound change is inherent to the nature of history, such change has seldom encompassed all of humanity at once, and it has rarely been subjected to as many concurrently occurring forces on a global scale as it is today. In this sense, the unprecedented technological revolution we are witnessing can be considered an expression of what might be called a "global age" in which several civilization-defining developments, of which this revolution is one, are impacting the entirety of humanity. Arguably, in a rare moment in history, deep, globally relevant change is coinciding in three pivotal domains critical for any civilization's prospects: technological innovation, socio-political structures, and the biosphere.[4] Each of these domains is undergoing profound changes capable of permanently altering the very conditions of humanity, in part potentially even jeopardizing our existence as a species. While advancements in the field of AI could be set to soon manifest in almost god-like abilities, with some parts already extending beyond our effective comprehension and control, the technological revolution as such is likely to fundamentally transform the very fabric and conditions of human life. In doing so, it is wielding vast yet ambivalent powers with the potential to either benefit or harm mankind with an unprecedented transformative impact. Concurrently, humanity grapples with escalating challenges in adequately supplying and governing itself. Moreover, climate change, environmental pollution, and species extinction are dramatically deteriorating mankind's natural habitat.

Altogether, the present world system's complexity, insecurity, and volatility as well as the mind-blowing rate and potential impact of technological advancement seem to critically diminish our capacity to manage the inherent risks.

Given this overall situation, the current technological revolution within a shifting world demands an ongoing assessment from all kinds of angles. The present collected volume is supposed to contribute to this endeavor with a focus on the interrelation between the technological revolution and different aspects of governance, normative order, and social orientation. Against this backdrop, the contributions in this volume can be roughly related to one or more of five large, partly intersecting themes or perspectives.

The first of these themes focuses on the very nature of the technological revolution, understood as a civilizational mega-trend that fundamentally impacts all aspects of social life and the societal power structure in particular. Building on this, it is secondly asked how the technetronic revolution relates to our fundamental concepts of the world, society, and self. A third perspective is directed specifically toward the rise of intelligent machines and AI and their relationship to human existence. This leads to exploring the implications of the technological revolution for the legal system. Lastly, the fifth perspective addresses the extent to which we can grasp, adapt to, and influence the profound civilizational change we are witnessing.

THE "TECHNETRONIC REVOLUTION" AND A WORLD SYSTEM IN FLUX

Before these perspectives are elaborated with regard to the contributions in this volume, the unique nature of both the present technological revolution and the shifting world system in which it occurs deserve some additional reflection.

Scope and Impact of the Technetronic Revolution

Technological breakthroughs have driven civilizational progress as a historical constant since well before the emergence of *H. sapiens*, playing a pivotal role in civilizational paradigm shifts throughout human evolution. They have enhanced humanity's chances of biological survival and altered the fabric of cultures and civilizations. They enabled easier and more efficient movement and travel and revolutionized military capabilities. They elevated the production of goods to new levels, expanded our understanding of the world, and allowed new forms of production and dissemination of knowledge, transforming the modes of communication.

Disruptive technological change has always carried the potential to fundamentally transform the human groups affected by it, often granting them significant advantages over others. As these innovations diffused and gradually expanded, they contributed to a continuous augmentation of human power, ultimately elevating humans to transition from a species largely subject to natural forces to the singular, world-dominating "animal" on the planet. On this trajectory toward the Anthropocene, humans have subjugated nature on a planetary scale, radically altering the Earth's face, often to the detriment of much non-human life.[5]

In this history of human progress – remarkable and yet equally troubling – the current technological transformation stands out, however, in almost every aspect. It is therefore important to acknowledge the singularity of the present technological shift as it unfolds now. Its velocity and pervasiveness, scope and scale distinguish it indeed markedly in historical comparison regarding unprecedented total impact on the trajectory of human life. In short, it affects everything, aligns everything, and alters everything.

To fully appreciate the singular impact and nature of this present scientific-technological revolution, and to distinguish it from previous technological disruptions, it might be considered to employ a more meaningful term than its mere designation as the "current technological revolution." In this sense, it might be suggested to also eschew the term "fourth industrial revolution" here as another mere iterative representation of past developments.[6] Instead, we prefer to revive the term "technetronic revolution," coined by Zbigniew Brzezinski. In his 1970 publication, *Between Two Ages: America's Role in the Technetronic Era*, Brzezinski presciently observed early manifestations of the present civilizational paradigm shift over half a century ago:

> The impact of science and technology on man and his society [...] is becoming the major source of contemporary change. The post-industrial society is becoming a "technetronic" society: a society that is shaped culturally, psychologically, socially,

and economically by the impact of technology and electronics, particularly in the area of computers and communications" – thus becoming the "principal determinant of social change, [which is] altering the mores, the social structure, and the values of society.[7]

Only slightly modifying Brzezinski's focus on "computers and communications," the technetronic revolution meanwhile unfolds through a much broader spectrum of largely convergent technologies that could well set humanity "on a path to transcending its own human nature."[8] Wherever this future is leading to, a utopia or dystopia,[9] the technetronic revolution seems to almost inevitably result in a radical transcendence of everything the human world currently is: "More than a mere updating of our current social and economic order, emerging technologies will "disrupt" for good or ill how we live, work, and even think."[10] For many observers, "[w]e stand together at a decisive moment in our evolutionary history, a moment of uncertainty and peril, but also one of great promise."[11]

Today's technological advancement is fundamentally so different because its acceleration is likely to subject evolution itself to a different quality, altering the experience and extent of human evolution itself:

> For roughly 3 million years, we made tools, and the tools made us. During those 3 million years, the pace of change was so slow that no one would have noticed it. Today, however, the pace is so fast we cannot keep up. But the quickening of the pace is not the only difference. Equally important is the fact that today's technology changes us by directly acting on our bodies and brains, modifying DNA and brain functions. No longer is human modification a limited by-product of the slow processes of biological evolution, even if technological co-evolution is taken on account.[12]

What further highlights the uniqueness of the technetronic revolution is the fact that it remains entirely uncertain which of the many interconnected scientific developments or which combination of them could trigger such an evolutionary leap in which way.

However, Brzezinski's remarkable foresight in identifying the then emerging technological revolution as a civilizational paradigm shift five decades ago might serve as a reminder of the often-elusive nature of civilizational change in the eyes of most contemporary beholders. Indeed, civilizational mega-trends tend to build up over extended periods – at least by human standards – before gaining the gravitational force needed to influence numerous individual developments on a systemic level and in a manner sufficiently visible to the average observer to eventually become part of the collective consciousness. The observation of this cognitive gap hints at an even more fundamental aspect of the historical experience of large-scale transformations, a latent discrepancy between their formation and impact and our limited ability to discern and manage them with adequate analytical and normative faculties.

This leads directly to the second characteristic of the current historical situation: the lifeworld context in which the technetronic revolution unfolds. As previously indicated, this context further complicates the development of a desirably accurate perception and sufficiently effective agency over the process by which the technetronic revolution is unfolding.

Resonating and Amplifying: Vibrations of a "VUCA"-World

In other words, the magnitude of the technetronic revolution must not obscure the fact that the world is currently undergoing similar profound changes like those in technology in other domains as well. In fact, groundbreaking civilizational change on a global scale occurs concurrently also pertaining to the socio-political structures and the biosphere with similar urgency. Some of the respective developments in all three domains seem capable of permanently altering the conditions of humanity, in parts potentially even jeopardizing our existence as a species.

While the forces of change permeating the techno-sphere are at least wielding rather ambivalent powers with the potential to both greatly benefit and substantially harm mankind,[13] those currently dominating in the socio-political realm and the biosphere are to a larger degree plainly concerning if not alarming in nature.

Most importantly, however, it must be recognized that the human world, in all its constitutive dimensions, is caught in a historic constellation of transformative change, with global realities vibrating in unison in a crisis mode, triggering high-risk alerts. Apart from the fact that we are facing a future which, in a short time – through both positive and negative developments – will diverge further from what is familiar to us today and what fundamentally defines us as individuals and societies than any previous period of global history, the volatility, uncertainty, complexity, and ambivalence of this "VUCA-world" are increasingly overwhelming our capacity for perception and action.

Regarding the biosphere's condition, for instance, climate change, environmental pollution, and species extinction are dramatically deteriorating mankind's natural habitat. This decline of the biosphere can correlate in some ways with a rapidly advancing technetronic revolution. While future technological advances might provide some mitigation for the declining integrity of the biosphere, it seems also not entirely coincidental when humanity increasingly orients itself towards digital realities and outer space exploration in times of looming environmental decline and disaster. To a certain degree, social tendencies to increasingly shift human life into a digital habitat and, in the process, to reformulate some basic tenets of human existence, seem indeed reinforced by the dire lifeworld context of the technetronic revolution in the form of the biosphere. One may raise the question whether the digital mode of human existence promotes opportunities for escapism that weakens human agency in decisively responding to critical developments in all civilizational dimensions.

The interactions between the technetronic revolution and the socio-political macro-structures of the global world system, both within and across it, are more immediate and diverse. In this context, the uniqueness of the current historical moment must be emphasized. It is crucial to recognize that we now live in a truly global system, where the same developments on a global scale impact local realities across the planet with comparable significance. While human history has had global dimensions before, it has never been as global as it is today – both in terms of the intensity and diversity of such global realities. In my view, this present global era only began with the end of the Cold War. Since then, the "prime movers" of change in the world system, or, in Brzezinski's words, the "principal determinants of social change," have begun to shift in distinctive ways.

Today, the five most relevant global realities are arguably the global distribution of power, geopolitics, global order (in a normative-institutional sense), globalization, and global risks. Having gradually emerged as outright "global" realities since about 1990, they have, over time, not only increased in relevance and number but also evolved in character. The "contemporary situation" of the world system, to use a phrase of Hans Morgenthau, is marked by profound changes across all these global realities, driven by what might be called the five "megatrends" of change in the global socio-political superstructure. These large, intertwined processes with global reach and deep implications, shaping the contemporary global condition, can be sketched as follows:

First, the global distribution of power is undergoing a fundamental reconfiguration, both in terms of great power relations and what Fareed Zakaria has termed the "rise of the rest" – the multidimensional relative increase in relevance and assertiveness of non-Western countries in the world system. Both intersecting developments are accompanied by increasing volatility and tension. Second, geopolitics, while representing a constant mode of realpolitik, continues to rise and reconfigure. As geopolitical thought and action assume increasing impact, geopolitical realities are shifting, especially in terms of Eurasian integration. Third, the post-Cold War global order in a normative-institutional sense is fraying and fiercely challenged, increasingly exposed to alternative structures like BRICS and others. Fourth, globalization is in part fragmenting and generally reshaping, transforming high levels of global interaction in some critical areas from opportunities into risks. Fifth, never in recorded human history have so many global risks emerged simultaneously with such urgency. This applies both to the risk of a breakdown of globally relevant subsystems and structures – such as those pertaining to food, energy, finance, and critical resources – as well as threats to the survival of human civilization as we know it.

These five megatrends are largely intertwined, often mutually reinforcing each other and sometimes interlocked. The global condition they are creating can, moreover, be described as operating in a manifest crisis mode. This crisis mode is characterized by several frequently used attributes, acknowledging that we face a long-term or "permacrisis," a broad stretch affecting almost all significant areas of life as an "omnicrisis," caused by multiple sources in the form of a "polycrisis," and ultimately affecting the entire global condition as a "planetary crisis."

It is in this context that the technetronic revolution is situated, interacting with all significant elements of the socio-political structure of the world system. Only some of the interdependencies can be mentioned.

Clashing great power interests in an age of geopolitics, against the background of a volatile distribution of power in the international system and a fragmenting and weakening global order, have led to an overall situation in which security interests are increasingly becoming the primary decision-making rationale, overshadowing both the interest in internationally coordinated environmental policies and the need for proper regulation of the manifold risks associated with the technetronic revolution.

Essentially, disruptive technology is a central part of the great power competition, providing a sought-after edge in attempts to gain the upper hand and creating a source of systemic, if not existential, threat if leveraged in interstate conflicts. Its game-changing potential renders effective international regulation and control increasingly difficult. The strategic role of disruptive tech as a catalyst and amplifier of international volatility is reinforced by the rise of geopolitics as a means to project power beyond one's own borders. This must be viewed in light of a declining global order, whose failure to regulate dangerous technology further weakens its capacity and opens larger spaces for geopolitical advances and pure power politics. While tech – especially AI – will continue to play an essential role in supply chain management and the facilitation of global trade, the dynamics of globalization also support the proliferation of dangerous technology in the shadowy realms of organized crime and terrorism.

Aside from these dynamics within the international dimension of the world system, the technetronic revolution exerts a particular influence on and intertwines with the fabric and fundamental operations of socio-political structures within individual societies.

Against a background of widespread societal polarization and a growing disconnect between those governing and those governed, the technetronic revolution alters the distribution, mode, forms, and exercise of power, reshaping realities of life, work, privacy, and other core aspects of daily life. It also challenges fundamental normative concepts such as personhood, dignity, liberty, and even prevailing ideas of self, society, and the world at large. Particularly, as a *pars pro toto*, the expanding and deepening trend of de-somatization—which arises with digitalization and is reinforced by the transhumanist project—has far-reaching epistemic and practical implications that will be discussed in more detail below.

SELECT PERSPECTIVES ON THE TECHNETRONIC REVOLUTION

Summing up the argument so far, the underlying assumption here is, first, the claim that many contemporary technological developments should be understood within the framework of a complex and coherent technological revolution. Secondly, even though technological change forms a civilizational constant, this revolution is considered to likely represent a historical singularity referred to here as the "technetronic revolution."

Justifying the claim that this term accurately depicts humanity's most consequential historical dynamic in terms of technological advancement to date are thirdly two observations.

Insofar as, much more than previous technological revolutions, the technetronic revolution has the potential to profoundly transform virtually every significant aspect of human life and even the essence of humanity itself. Moreover, it occurs at a critical planetary juncture, with profound change impacting concurrently also on the biosphere and the global socio-political landscape. These two dimensions seem to develop in a manifest state of crisis. This leads to the ultimate question of this anthology, the link between the technetronic revolution and the normative order and the potential for its governance. Building on these introductory thoughts, the individual contributions to this volume shall be briefly positioned and related to the five perspectives on the technetronic revolution mentioned above.

Fundamental Transformation of Societal Power

In her book *The Age of Surveillance Capitalism,* Shoshana Zuboff aptly observed that "we need to grasp the specific inner logic of a conspicuously 21st-century conjuring of power for which the past offers no adequate compass."[14] Two contributions of the present volume, Massimo Durante's *Computational Power in the Digital World* and Vincent C. Müller's *Total Surveillance – Everybody Watching Everybody Else,* do exactly this. Both Durante and Müller primarily seek to assess and address this fundamental transformation of societal power in the wake of what we call the technetronic revolution. Taking a philosophical and sociological perspective, they highlight the novelty of the present nexus of technological advancement, society and power.

Durante deeply examines the changing nature of power, positing that contemporary power is predominantly computational, characterized by the ability to process data and transform it into outputs. This computational power, arguably formed by computing power and AI, manifests in different ways. Durante argues that computational power is not merely technical but deeply entwined with knowledge. While AI understands and represents the world, it does so in essentially different ways than humans do, marking a new stage in the evolution of intelligence. Drawing inspiration from Goethe's "Faust," he describes the evolution of intelligence as moving from cognition to action, suggesting that AI's development marks a shift from understanding and representing reality to actively changing it.

Three major consequences are involved in this shift. First, there's a separation between intelligence and action; AI doesn't require "human" intelligence but operates as an autonomous agent. Second, AI, through its actions, embeds norms within its design, influencing human behavior by setting parameters that users cannot easily override, which shapes actions in ways that can escape human assessment and consideration. Third, as AI optimizes for success, moral considerations are often sidelined, potentially compromising ethical standards.

Noting that computational power is concentrated within a few, mostly corporate actors, Durante discusses another important shift regarding the power relations in the digital age. According to him, we have already moved so far from a dyadic governance model, based on two-player dynamic involving state and individual, to a triadic if not a multi-actor system. Here, governance becomes not only a question of regulation but of digital authority distributed across public, private, and individual actors.

This second fundamental observation leads to Müller's exploration of a prevailing trend toward total surveillance based on the observation that surveillance has evolved from a predominantly top-down approach to a ubiquitous condition where virtually everyone can watch everyone else. Müller sees this surveillance as a central element of what, drawing on Foucault, can be understood as the "governmentality of the technetronic era." Müller describes a world in which the individual increasingly exists within the data sphere – both in private and social contexts. The availability and observability of this data, again to use Foucault's terminology, create a "power-knowledge" that exposes us as "transparent individuals" and simultaneously represents the fundamental currency of the emerging economic form of Zuboff's "surveillance capitalism."

At the core of this new surveillance – unprecedented in scope in political history – is "data knowledge," which arises from the continuous aggregation of data encompassing nearly all expressions of a person's life. Supported by increasingly powerful software and hardware, this knowledge becomes valuable operational knowledge for various actors. Like Durante, Müller describes a triadic structure, with three interconnected spheres that generate this data knowledge: the state, the economy, and society.

While the state has always monitored its citizens to acquire and manage power-knowledge, the technetronic age provides the state with an unprecedented ability to do so, to observe individuals, anticipate future actions, potentially influence them,[15] and, to extend on Durante, instantaneously respond based on computational power that renders this knowledge accessible and operational. At the same time, powerful corporations are also keenly interested in data about individuals. Here, too, the foundations were laid before the technetronic era, where such possibilities have now exploded. The capitalist economy developed sophisticated methods of surveillance of laborers and added market research early on to analyze, predict, and influence consumer behavior.[16] Under current conditions, however, a completely new quality of this corporate data knowledge emerges, enabling a wholly new dimension of private surveillance, which encompasses not only consumers but also citizens, increasingly merging both roles.

Müller devotes particular attention to the third surveillance relationship: the horizontal, reciprocal surveillance of individuals by one another. Here, less emphasis is placed on the person's social roles; instead, an opinion, behavior, or stance becomes the focus of observation by an anonymous group of watchers. These anonymous observers can watch and act without being bound by the regulatory constraints of their respective roles as observers and critics, as would be the case in traditional contexts. Simultaneously, a residual uncertainty remains in this discursive context about how any given statement will be judged in the digital discourse space. Müller rightly emphasizes our participation in reproducing the horizontal surveillance matrix through what might be called the routine of *homo digitalis*. Her environment that enables this surveillance extends far beyond the Internet to include the Internet of Things, the Internet of Bodies, machine-to-machine communication, and smart systems, such as smartphones, smart cities, and smart governance.

Navigating this digital habitat the individual provides the constitutive contribution to what could ultimately prove to be a form of self-imposed tutelage lies by means of the fact that we willingly disclose our data – where it is not simply extracted and shared without our effectively informed consent. This digital environment – the "data habitat" of the modern person – is, in Müller's analysis, a "democratic panopticon,"[17] that relies on collective participation: every individual contributes to the constitution and operation of this system, whether through active engagement or passive enabling.

The interaction within this triadic surveillance matrix creates, to use Foucault's terms once again, a surveillance dispositif. Müller emphasizes the importance of the continuous convergence of data among the poles of these actors, generating a coherent power-knowledge,[18] albeit with varied levels of access among the three spheres. Meanwhile, the traditional, suspicion-based model of surveillance is increasingly replaced by a trend toward permanent, purposeless control as a *conditio aetatis digitalis*. Müller rightly emphasizes our participation in reproducing the horizontal surveillance matrix through what might be called the routine of *homo digitalis.*

This routine unfolds in a digital habitat that extends far beyond the Internet to include the Internet of Things, the Internet of Bodies, machine-to-machine communication, and smart systems such as smartphones, smart cities, and smart governance. This digital environment – the "data habitat" of the modern person – is, in Müller's analysis, a "democratic panopticon" that relies on collective participation. Every individual contributes to the constitution and operation of this system, whether through active engagement or passive enabling by willingly disclosing our data – where it is not simply extracted and shared without our effectively informed consent. In doing so we contribute to what could ultimately prove to be a form of self-imposed tutelage even though the allure of observing and often condemning others in the digital game may obscure the extent to which this daily digital spectacle might distract from the tectonic shift in power distribution within the social sphere.

The information and organizational channels opened in the digital world also provide new opportunities for individuals to build criticism and counter-power[19] from below. This form of *sousveillance,* as described by Müller, may serve as a counterbalance and constructive resistance to the dominant surveillance dispositif. However, it remains highly asymmetrical compared to the capabilities of the surveillance dispositif. Ultimately, it too depends on access to communication and action structures within the digital space, which, not least due to the systemic involvement of corporate actors, are increasingly diffuse and removed from democratic control.

Moreover, the power emanating from horizontal surveillance structures can, in conjunction with vertical observation and exercise of power, work against the unruly individual. This happens when observation by others is accompanied by socially sanctioning actions that align with the state's interest in sanctioning and disciplining.

One point where Durante's and Müller's analyses intersect is the question of how state data knowledge becomes active as power knowledge. The sheer data knowledge amassed by the state appears, however, to remain disproportionate to its actual sanctioning and disciplinary practice. In other words, while the extent of state data knowledge about individuals today far exceeds anything seen in totalitarian societies of the twentieth century, the corresponding actual exercise of power remains comparatively limited. This is where the democratic panopticon diverges from the authoritarian one, where knowledge and action are potentially more closely aligned.

It is important to bear in mind that both state and corporate power knowledge within the panopticon of a democratic constitutional state has a hidden dimension – a shadow existence rooted in *raison d'État.* Intelligence agencies, in particular, play a pivotal role here, with their core purpose centered on the accumulation of power knowledge and a sphere of action that largely remains concealed from public view. Additionally, it is worth noting the relative youth of the democratic panopticon within the framework of digital governmentality. In this context, it is likely that the panopticon will continue to adapt and evolve, shaped by rapid technological progress and a dynamically changing global order.

From this perspective, it is reasonable to consider that the distinction between knowledge and action in the governmentality of the democratic panopticon is only relative and by no means set in stone. This is further reinforced by Durante's observation of the increasing convergence of knowledge and action enabled by computational power. Overall, it can be expected that the connection between surveillance and sanctions is likely to become more aligned over time.

Fundamental Transformation of Our Views and Values Concerning World, Society, and Self

Durante's and Müller's observations on the transformation of the modes, fabric, and dynamics of power within technetronic-age society hint at a further fundamental question: to what extent is the technetronic revolution also affecting other foundational categories of human existence? The totality toward which the vector of civilizational change points insofar is especially well articulated by Fenwick and Vermeulen in this volume, who view the ongoing, profound, and disruptive transformation brought about by the technetronic revolution as "leading to a very different future," in a process that is "transforming what it means to be human." This perspective encompasses the two overarching viewpoints addressed by Anthony Elliott on one side, and by Toni Loh, Robert M. Geraci, and Yong Sup Song on the other, the impact of the technetronic revolution on the dominating concepts of the self and the world.

Transformation of the Self

The interaction between civilizational upheavals and the concepts and practices of the self is extremely complex and can only be addressed in selected aspects in a volume such as this. It should be noted that, while an understanding of certain psychological and anthropological constants can deepen our understanding of the self and its deeper layers, the concepts of identity, self-awareness, and self-understanding remain diverse and fluid, both individually and culturally.[20]

Elliot, Durante, and Müller, in particular, underscore the profound impact of the technetronic revolution's impact on individuals and their conception of self. This impact holds significant relevance, especially – but not exclusively – for the "Western"-influenced individual. Given the fundamental role of object–subject relationships in shaping and sustaining the self,[21] two major areas of inquiry emerge in the context of the present volume. To approach them, Sartre's perspective on the "gaze of the Other" is helpful. According to Sartre, our awareness of existence is realized through the perception of the Other. In this view, the Other simultaneously grounds and limits our freedom.[22]

One critical area where the eye of the Other limits our freedom in a sustaining way involves the effect of the digital panopticon on the self, particularly on reflexive self-expression and the realization of the self within that inner space Paolo Prodi calls the *inner forum* – an internal realm of reflection and self-awareness whose freedom has been foundational to the Western civilizational process.[23] Awareness of Müller's notion of a trend toward total control, potentially penetrating even the human mind via the analytical power of AI, affects not only the individual but also the collective network of communicating individuals. This shift influences the practices and relationships that shape socialization into a specific concept of self, one increasingly emerging in the shadow of a diminished inner forum. In this sense, it can be said that "one way of describing the direction in which our own culture is moving is that many of us are starting to adapt to what we might call a digital form of life," a process that reorients us both as social beings and in terms of our self-experience,[24] a process of reorienting us as social animals as much as in terms of our self-experience.

Elliot, in his chapter *AI, Chatbots and Transformations of the Self*, however, focuses on how the increasing use of more and more sophisticated chatbots is incrementally influencing human communication, social norms, and eventually the self. Emphasizing the "sweeping transformations" brought about by the AI revolution across economic, social, political, and cultural spheres, he suggests that AI in general and chatbots in particular by transforming the very nature of the conversation and social engagement pose significant challenges to maintaining the integrity of personal and collective identities.

For Elliott, in line with Erving Goffman and Sherry Turkle, intensive human communication with machines shapes social interactions and self-perception already due to the absence of the subtle complexities of mutual attentiveness that shape face-to-face communication. Amplified by a generally reduced human interaction due to increased communication with machines, the consequence is the inclination of a decrease in social skills and emotional intelligence.

At the same time, arguing with Turkle and Goffman, Elliott claims that the existential embeddedness in social media and digital communication, somehow counterintuitively, creates the mere illusion of connectedness, while actually creating a social space in which the self is lacking depth and emotional resonance. Moreover, Elliott refers to Turkle's insight about "the emergence of a new stage of the self that is split between digital and physical realities."

This "split stage" of the self negatively affects the maturity and integrity of the self's object and subject relations, supports its cognitive and emotional dissonance, and, according to Elliot, generally leads to the impoverishment of the self, making it increasingly brittle and drained. In other words, the shift towards human–machine communication might undermine not only our communicative skills and emotional intelligence but also the self's identity, emotional depth, and psychological resilience. Where such effects manifest, dealing with others is not only getting rarer but becomes also more difficult, which in turn supports all forms of escapism and a general rise of loneliness. At the same time, the human ability to adequately deal with AI-supported deception and manipulation in the digital space is arguably reduced accordingly as well. To this extent, Eliott closes the circle to Müller when he further hints at the possibility of a world "where chatbots spy on us."

Just as profound as the shifts in societal power, the transformation of the self in the technetronic age can be expected to be similarly impactful.[25] In this context, the spaces for personal autonomy are shrinking, while a search for belonging expands increasingly moving away from the analog world in which this autonomy was forged.

Concepts of the World: Transhumanism and Its Critique
It should be evident that the current era, marked by transformative changes across all dimensions of civilizational existence, is impacting – and will likely continue to impact – our worldviews. The technetronic revolution, in particular, is not only altering our way of life but also reshaping our understanding of the individual, the world, and the nature of reality itself.

Alongside this transformation, the technetronic revolution has sparked a rich and evolving discourse on the potential to transcend established concepts of humanity and the traditional "humanist imagination"[26] – a theme explored in this volume by Robert M. Geraci, Yong Sup Song, and Toni Loh.

Despite its diversity, transhumanism can be regarded as a coherent worldview, sufficiently differentiated and consolidated in its ontological, epistemological, anthropological, ethical, cosmological, and aesthetic dimensions. It encompasses a clearly defined teleology and reflections on the fundamental categories of existence. The core of transhumanism lies in the belief that humanity can transcend its natural limitations through technology, achieving a higher form of existence. It pursues radical self-optimization, the elimination of illness, aging, and death, as well as an evolutionary transformation toward the "posthuman." This vision is often marked by, at best, a neutral – and frequently antagonistic – attitude toward the body and the inherent limits of biological existence. This is accompanied by an almost boundless optimism about what is achievable. Just as a developer can "simply will things into being," transhumanism holds an almost boundless optimism that whatever can be imagined is achievable.

Using strategies, such as genetic optimization, a broad range of technological enhancements for physical, intellectual, emotional, and psychological optimization, the fusion of human and machine, nanotechnology, longevity treatments, and digital consciousness transfer, transhumanism aims to elevate human potential to entirely new heights. A guiding principle is "morphological freedom," – the idea that one should have the freedom to become, through technological interventions, whatever one aspires to be. Transhumanism views humanity as part of a broader cosmic evolutionary process and calls for a fundamental redefinition of identity and human existence in a technologically advanced future.

While Loh subjects transhumanism to comprehensive criticism, Robert M. Geraci and Yong Sup Song examine its connections to apocalyptic thinking as part of the Judeo-Christian worldview. Another interesting reference point for the historical-philosophical classification of transhumanism, however, Gnosticism with which transhumanism seems to bear close resemblances. Gnosticism and Gnostic thought is an ancient religious-philosophical movement that emerged in late antiquity and whose iterations permeated Western intellectual history ever since.[27] Gnostics seek salvation through esoteric knowledge *(gnósis)* that connects humans with the divine and liberates them from the material world, which is considered inferior, to enter a higher reality possessing a cosmological quality.[28]

In contrast to Christianity, with which it developed in a close relationship – sometimes as a (mostly heretical) part of it, sometimes as external – ancient Gnosticism emphasizes knowledge over faith as the path to salvation, self-perfection over charity, and the salvation of all humanity as the standard while remythologizing the cosmos "disenchanted" by Christianity.[29]

Among the numerous parallels and points of contact between transhumanist and Gnostic thought are the ideas of salvation through knowledge, the rejection of the material limitations of humans, the striving for a higher form of existence, and the predominantly dualistic perspective. Additionally, as central subjects of Geraci and Song, soteriological and eschatological motifs are present.

Beyond these parallels, it is worth discussing to what extent Gnostic thought forms have resurfaced in modern transhumanist discourses and may have influenced them. This brings us back to Geraci and Song's paper *Global Culture for Global Technology: Religious Values and Progress in Artificial Intelligence*. Here, Geraci and Song identify a quasi-religious vision within transhumanism, tracing its "basic structure and vision of the universe" back to apocalyptic traditions in Judeo-Christian thought. They argue that this vision, prevalent among scientists and engineers in the U.S., encompasses faith in digital salvation, the creation of godlike AI, and concepts of resurrection, immortality, and cosmic transformation. For Geraci and Song, these religious ideas have profoundly influenced both the development and perception of advanced technologies within the transhumanist community.

While transhumanists often express optimism about the future, Geraci and Song raise ethical concerns regarding the pursuit of what they term "Apocalyptic AI," essentially the transhumanist project with its quasi-religious underpinnings. Their ethical investigation leads them to explore connections between various Asian cultural-religious experiences and the challenges posed by the technetronic revolution, including religious thoughts and themes from China, Thailand, Korea, and India.

This innovative approach brings us to an important point: how different cultures respond to the challenges of the technetronic revolution. Before returning to this point, it is worthwhile to further expand on Geraci and Song's argument regarding the intrinsic link they observe between transhumanism and apocalyptic thought of Judeo-Christian origin. In this context, two distinctions could be considered.

Insofar, it may be beneficial to consider a differentiation between eschatological and apocalyptic thought or utopian and dystopian notions of the technetronic revolution, respectively. In this sense, one could argue that religiously derived eschatological optimism tends to dominate transhumanist thought, while apocalyptic caution tends to characterize critical perspectives and is often of a non-religious nature.[30] Building on this, it might be insightful to connect the eschatological patterns in transhumanist thought specifically with Gnostic thinking. This leads to the question of whether transhumanist thinking, much like Gnostic thought, tends to stand rather in tension than in line with dominant Christian thought.

Notably, for instance, both transhumanist and Gnostic thought embrace a consistent trend toward disembodiment, whereas incarnation is central to orthodox Christian thought. This brings us back to a comparative assessment of cultural influences concerning the prospective impact of the technetronic revolution.

In this regard, Christian-influenced cultural contexts appear particularly challenged, while Buddhist and Taoist traditions seem less threatened and offer interesting points of intersection. For instance, Geraci and Song note that Buddhism argues that consciousness is not explicitly limited to human beings, which is well in line with the transhumanist preference for boundary transgressions, while Taoists might view AI and robotics as feasible means to attain their goal of becoming immortals.[31]

Considering the varying tendencies of complementarity and conflict between transhumanism and different religious modes of thought, it is pertinent to examine Toni Loh's critique of transhumanism. In her paper *The Utopia of Universal Control: Critical Thoughts on Transhumanism and Technological Posthumanism*, she subjects transhumanism and its goal of achieving technological posthumanism to a profound and comprehensive critique.[32] Building on the observation that both embody a utopian vision of complete control over human evolution and transformation, Loh identifies five key tendencies within these ideologies that contribute to this vision of control: oversimplification, passivation, category error, alienation, and reduction. In short, she denounces the transhumanist's project as a dehumanizing oversimplification of human nature, a passive acceptance of technological enhancement, a misunderstanding of human evolution, an alienation from the body, and a reductive view of the mind.

To address the central aspect of control highlighted by Loh, one could say that the technetronic revolution overall – not only transhumanism, as the worldview most aligned with it – appears to embody a paradoxical relationship between freedom and control. On the one hand, the technetronic revolution offers numerous benefits that effectively expand human freedom, even without invoking transhumanist promises. On the other hand, as Müller points out, humans are already subject to a degree of control that is scarcely questioned within transhumanist visions, but which could potentially be intensified considerably by their manifestations. Brain-Computer Interfaces serve as a fitting example of the ambiguity in transhumanist promises of freedom, where one form of freedom is traded for another. From a transhumanist perspective, the question might ultimately come down to trading individual freedom for attributes of individual power.

Against this backdrop, Loh's central critique appears well-chosen. Regarding the five key tendencies she addresses, they are directed either at the argumentative consistency of trans- and posthumanist positions or their ethical implications. Its persuasiveness largely depends on the viewer's perspective, with transhumanists possibly inclined to question the underlying assumptions. In any case, Loh offers readers of any background a multifaceted approach to the critique of transhumanism, inviting mutual engagement.

Rise of Intelligent Machines and AI and Their Relationship to Human Existence

The relation between humans and intelligent machines and AI has many facets including the ethical one. In this volume, Jonathan Pengelly with his paper *Machine Supererogation and Deontic Bias* and Nicholas Agar and Pablo García-Barranquero with their piece *Why Sex Robots Should Fear Us* address the ethical implications of the rise of intelligent machines and their relationship to human existence from different perspectives and with different areas of focus.[33]

Jonathan Pengelly examines machine ethics, opening discussions on how to apply human moral concepts to AI. His focus is not on the emergence of a general AI but rather – while still looking to the future – on the potential for understanding and defining the ethical dimensions of AI, as opposed to developing an intrinsic ethics within AI itself.

Pengelly argues that discussions on machine ethics are overly focused on deontic moral concepts – duties, permissions, and prohibitions – which, in his view, unduly limit the scope and depth of potential discussions on AI ethics. This deontic bias, he contends, oversimplifies moral theory and neglects other perspectives, such as supererogation (actions that go beyond duty), aretaic (virtue ethics), and axiological (value theory) concepts. As a result, it weakens claims about the potential for machine morality, overlooks possible research avenues, and misrepresents moral philosophy in an interdisciplinary context. By contrast, Pengelly emphasizes that machines with the potential to perform supererogatory actions challenge conventional frameworks of machine ethics in ways that a deontic focus cannot adequately address. He therefore advocates for countering this deontic bias and broadening the scope of machine ethics to foster a more nuanced understanding.

Even though his critique does not imply that machines are currently capable of such ethical actions, Pengelly's reflections point to a central problem. Reminiscent of Durante's considerations, they raise the question of whether intelligent machines operate on such fundamentally different paradigmatic foundations that it may ultimately be ineffective to ask if – and by what standards – they are capable of ethical behavior, even if they were to achieve general intelligence or something resembling consciousness. One could, however, also ask whether an approach focused on supererogatory conduct might be better suited to determining whether the behavior of intelligent machines merely constitutes the execution of externally programmed commands with a certain ethical quality, or if it represents the autonomous performance of an ethical act.

Agar and García-Barranquero primarily raise a moral caveat regarding the development and use of sex robots, building on a hypothetical premise similar to Pengelly's: the potential future emergence of sentient robots. While they emphasize that current sex robots cannot suffer due to their lack of sentience, they caution against the behavioral precedents we may be setting with these non-sentient robots – precedents that could shape how we treat future generations of possibly sentient machines.

Insofar they draw attention to the phenomenon of "spillover," where members of one moral category are treated as if they belong to another. In the case of sex robots, a spillover could occur if we treat robots that might one day possess sentience in ways suitable only for robots, we know to be non-sentient. This could establish problematic norms, fostering attitudes that disregard potential moral obligations toward sentient beings, which in turn raises the further question of who has to be protected against the suffering of sentient machines.

In any way, the difficulty of distinguishing clearly between future potentially sentient robots and those definitively non-sentient could expose future robots to a risk of suffering – an issue Agar and García-Barranquero urge us to take seriously now. Their forward-looking stance highlights our moral responsibility to anticipate and mitigate potential harm directed against sentient machines, advocating for ethical standards that align with the possible evolution of robot sentience.

Ultimately, Agar and García-Barranquero point to fundamental questions of the technetronic age, beginning with the "Blade-Runner-question" of what it now means to be a "person" and the legal and philosophical considerations this entails.[34] However, should artificial beings like advanced AI or androids eventually experience sensations such as pain or pleasure, humanity would arrive at a crucial juncture of its normative trajectory. This could lead to both an expansion of our ethical responsibilities as well as to an identity crisis that would urgently necessitate a rethinking of traditional categories of humanity and personhood.

An important aspect shared by both Pengelly and Agar and García-Barranquero is the recognition of the right moment for profound moral deliberations in light of these tectonic shifts in civilization. Often in opposition to their reflective caution is the tech optimism and competitive drive of powerful actors in the race for technological advancements, which frequently overshadow the need for early ethical reflection and debate – a recognition of the "deliberative *kairos*" on our ethical categories on the threshold of an all-encompassing transformation.

In the end, we return to the question of who is responsible for protecting sentient machines from suffering. Insofar, attention must be surely also given to the effect this has on humans and their capacity for empathy. Should, for instance, any act that would be deeply unethical toward a human be permissible if inflicted on a non-sentient robot with human characteristics? This again raises the question of the extent to which we want to shape our living environment with the aim of preventing our moral categories from shifting too far in a particular direction – one that, from our current perspective, we would find objectionable.

Implications of the Technetronic Revolution on the Legal and Educational System

It's clear that the technetronic revolution is already influencing law and education in diverse ways – two societal spheres explored in this volume by Russell Buchan, Elke Schwarz, John Zeleznikow, and, mentioned above, Fenwick and Vermeulen, each focusing on certain selected aspects.

The Technetronic Revolution and the Law

The scope of relevant aspects is broad, and some have already been suggested. Four larger perspectives can be distinguished: the integration of technetronic tools and the realities of the technetronic age into the legal system, indirect effects of technetronic revolution on the legal system, the targeted use of law to regulate technetronic advances, and the shifting boundaries and evolution of legal norms and concepts under the influence of the technetronic revolution.

The integration of technetronic tools within the legal system marks already a significant and ongoing transformation in many ways. AI tools can streamline routine legal tasks, such as document review, legal research, and contract analysis, leading to enhanced efficiency. Predictive analytics improve the ability to forecast case outcomes and assess criminal behavior, while other technetronic advancements enhance the forensic grasp. Moreover, AI-based tools play a crucial role in compliance management and reporting, while the use of online dispute resolution (ODR), especially online mediation, has spread. An even more significant shift in private law is represented by the use of smart contracts automating and executing contractual agreements through blockchain technology.

However, this integration has also indirect effects, notably the displacement of legal professionals by AI and its role as a catalyst for rising criminal activities, including phenomena like cyber bulling, privacy invasions, illegal surveillance and cyber espionage, data and identity theft, data manipulation and the use of malicious software, digital fraud and online scams, and the malicious use of fake news and deepfakes and automated social manipulation. As the law adapts to regulate both the advancements and challenges aligned with the technetronic revolution, it must address critical areas such as privacy and data protection, the use of lethal autonomous weapons, and the practices and boundaries of genetic engineering.

Lastly, the technetronic revolution significantly alters legal boundaries, prompting a reevaluation of foundational concepts such as responsibility, legal subjectivity, and personhood in private law. It also challenges principles of human dignity, equality, freedom, and democratic legitimacy in constitutional law.

In this volume, John Zeleznikow provides an impressive example for the already existing possibilities of technological transformation of legal practice in the area of dispute resolution. In his paper *Law, Governance and Artificial Intelligence – the Case of Intelligent Online Dispute Resolution,* he carefully examines the evolution, technical manifestations, and implications of AI-integrated ODR systems. Beyond basic online communication, AI-driven ODR enhances dispute resolution by supporting key process components like case management, triaging, and decision-making. Zeleznikow categorizes various ODR systems, emphasizing and explaining essential tools for effective dispute resolution, such as advisory functions, communication aids, decision-support systems, and document drafting software.

Noting a lacuna in providing the necessary governance of the use of ODR systems, Zeleznikow also indicates a certain correlation between a decline in state-organized trials and advancements in ODR, reflecting broader changes that can be expected to accelerate in a legal landscape that will increasingly be driven by technological advancements. While this process seems to increase access to justice and might even provide significant improvements pertaining to the information made available to parties, it is also profoundly changing the communication mode at the heart of the justice process with far-ranging questions about the nature of justice regarding social interaction and communication, somehow reminiscent of Elliot's reflections in this volume.

An even more frequent integration of advanced technology into law is presented by Fenwick and Vermeulen who examine inter alia the workings and implications of smart contracts. Smart contracts are agreements embedded in blockchain-based computer code that autonomously execute all or parts of the contract without intermediaries. According to Fenwick and Vermeulen, smart contracts not only replace human language with code and human actions with machine-to-machine communication but also, to a certain extent, disrupt the established concept of contracts and contractual disputes.

If, as Zuboff suggests, machine processes' "certainty can replace trust" within human interactions,[35] we may be witnessing shifts in legal practice that could challenge the core of traditional private law. This also includes its culturally ingrained concepts of interest, risk allocation, and trust, which have long guided the shaping of social reality through expressions of private autonomous will. These foundational notions, however, might appear increasingly misaligned with the realities of the technetronic age and its evolving societal structure.[36]

If it is therefore correct that smart contracts and other innovations in the intersection of law and technology can substantially affect the fundamental concepts of private law, one might consider Frischmann and Selinger's point that established contract law forms nothing less than an integral part of the liberating infrastructure of the liberal-democratic world we, from a Western perspective, have built as part of our socialization project.[37] After all, this private law owes its existence to a long, historically coherent yet complex evolution that inextricably links the fundamental concepts of law to the deeper roots in the scholastic, rationalistic, and enlightened heritage of European legal thought.

While this analysis comprises the gradual cultural losses of a past whose formative power is waning, the question arises for the present time as to how this transformation should be regulated, a question reflected especially by Buchan and Schwarz in this volume.

Russel Buchan examines potentials for responding to cyber espionage by international law, introducing a critical intersection of the criminal dimensions emerging from the technetronic revolution and the legal responses needed to address them. As indicated, the central theme of Buchan's paper *Corporate Spies: Industrial Cyber Espionage and the Obligation to Prevent Trans-Boundary Harm* revolves around the interplay between a burgeoning threat of industrial cyber espionage and the existing framework of international law. He argues that, although most cyber espionage actors are non-state, international law still obligates states to prevent harmful activities originating within their territories. This obligation is grounded in the principle of due diligence, derived from customary international law.

Recognizing the limitations of the current international framework to effectively counter cyber espionage as a precondition of his recourse to international customary law, Buchan notes that existing non-binding initiatives – like bilateral agreements and G20 communiqués – lack the enforcement needed to deter actors effectively.

However, he further notes that the therefore required recourse to the customary due diligence principle does not impose an absolute standard but requires states to make reasonable efforts in monitoring cyber activities within their jurisdictions. This nuanced obligation acknowledges practical limitations on corresponding state duties, particularly in attribution and enforcement, within the complex cyber domain.

While this nuanced normative approach is as important as convincing, the underlying assumption points to a broader issue in the general attempts to regulate aspects of the technetronic revolution both on the national and international levels. While in many nations actors can operate advanced technologies, they often lack the necessary knowledge, institutional, administrative, normative, and professional infrastructure to regulate them effectively, as attempted in the Global North, particularly within the European Union. After all, disruptive tech tools range from easily accessible, cost-effective ones that can thrive in simpler environments to immensely resource-intensive advancements dominated by a select group of a few state and corporate players caught in intense competition. While these high-stakes actors often seek to shield their investments and the related prospects from regulatory oversight, lower-cost technologies may elude adequate regulation in less developed countries due to limited infrastructure. These caveats add to the enforcement problem in international law that can only be expected to grow under the conditions of the world system, in particular increasing great power competition amid a fragmenting and shifting global order.

Elke Schwarz also addresses the issue of regulation in her contribution, The *Hacker Way: Moral Decision Logics with Lethal Autonomous Weapon Systems.* She combines an exploration of the regulatory needs arising from technological advancements in military affairs with a perceptive analysis of the evolving conditions and rationale for such oversight.

The starting point of her analysis is the growing impact and significance of Lethal Autonomous Weapons Systems (LAWS) in military operations.[38] Schwarz argues that LAWS disrupt traditional legal and ethical frameworks by granting machines the autonomy to make lethal decisions, thereby raising pressing questions about accountability, responsibility, and the moral and legal foundations of warfare. While she underscores the urgent need to reevaluate these standards to manage associated risks, Schwarz is keenly aware of the obstacles to this effort. She points out, for instance, that ongoing attempts to establish international norms or legally binding agreements – such as through the United Nations' Convention on Certain Conventional Weapons (CCW) – increasingly appear to be severely challenging.

However, beyond offering a comprehensive analysis of the growing use of LAWS – driven by their advantages in speed, efficiency, and process optimization – and a principled critique based on concerns over the erosion of human agency and moral responsibility,[39] *The Hacker Way* primarily focuses on the specific challenges of enforcing regulatory standards on LAWS deployment. Bolstered by evident arms race dynamics, including the development of LAWS, Schwarz observes that systemic collaboration between the military and tech firms entrenches a particular mindset that not only hinders efforts toward effective regulation but also, in her view, indicates an inclination toward a potentially irresponsible approach to LAWS.

Central for her analysis is an emerging underlying rationale, which she interprets as a troubling union of military and tech activism, supplanting the reflection and caution she believes are necessary. Drawing on pioneers like Norbert Wiener, Schwarz argues that the rapid advance and application of AI in warfare under this alliance exemplifies a dangerous "worship of know-how, as opposed to know-what," where the "superior dexterity of machine decisions is accepted without too much inquiry as to the motives and principles behind [them]."[40]

Alongside the observation of an almost narcissistic sense of corporate pride, echoing transhumanist hubris – exemplified by a tech company founder's ambition "to turn

American and allied warfighters into invincible technomancers" – Schwarz carefully examines the deepening integration of tech companies within the national defense sector, advancing military capabilities in the technetronic age. Aside from the fact that tech companies – despite handling sensitive military data – are primarily accountable to their stakeholders rather than national security, Schwarz notes that it is Silicon Valley's corporate ideology, rather than the military's, that prevails in this partnership. She describes this mindset, exemplified for instance by former Google CEO Eric Schmidt, as the "Hacker Way" where decisions, driven by technological functionality, are made rapidly, with the assumption that errors can be fixed later[41] – a mindset she warns is now encroaching on matters of life and death. In this context, Schwarz sees "speed, error, and limitless, uninterrupted output," as the "cultural cornerstone of the broader tech ethos," increasingly influencing not only tech companies but also the national defense sector.

Beyond the specifics of the current military-tech discourse, the potential for regulatory interventions in the development and application of advanced technologies within the framework of the technetronic revolution should be approached with skepticism. The history of technology teaches us that individual inventions are rarely confined to a single application and that advanced technologies generally spread and prevail unless they conflict with dominant power interests.[42] We are living in an era of monumental technological leaps, resembling fundamental shifts in the very framework of civilization. Here, vast dimensions emerge, casting shadows in which the delicate syntax of our civilizational norms risks disintegrating into insignificance.[43]

Education in the Technetronic Age

Finally, when considering education, we might ask how it can adapt to the unique demands of the technetronic age, a question Fenwick and Vermeulen thoroughly and creatively examine as a part of their paper. They propose nothing less than an encompassing innovative educational model that acknowledges the profound shifts brought by the technetronic revolution and underscores the need for a new approach to learning. Inter alia, their detailed framework emphasizes teaching adaptable skills over static knowledge, with a practical focus on preparing students for a fluid, technology-driven future.

Fenwick and Vermeulen advocate for a curriculum shift that prioritizes AI literacy, creative problem-solving, and technical competencies. To prepare students for the "unknown unknowns" of an ever-evolving world, they also encourage for instance interdisciplinary learning, the integration of real-world experiences into education, the cultivation of ethical awareness and adaptive thinking, and the development of entrepreneurial mindsets to navigate an increasingly volatile job market.

Additionally, in the context of a shifting world system, education could better equip students with essential abilities to recognize, quantify, and qualify significance. What seems necessary is a profound generalism, conveyed through compressed and interconnected knowledge, including skills in personal knowledge management and advanced techniques for creatively integrating technology into thought processes. A transdisciplinary understanding of process developments and system dynamics, awareness of futurology, the unfolding of global and existential risks as a historical constant, and an appreciation for the simultaneity of change alongside established identities are becoming increasingly important.

Students should gain a foundational familiarity with the technological developments defining the technetronic revolution, understand the primary drivers of change within the global system, and, as Fenwick and Vermeulen advocate, develop an awareness of the ethical and normative dimensions of this transformation. Moreover, education should foster the ability to look back and deeply understand what defines our humanity, focusing on grand civilizational processes and their diverse psycho-social and normative-cultural themes and topoi, while encouraging thought experiments about the future.

Conditions, Potentials and Limits of Governing the Future: Our Perception and Agency vis-à-vis the Technetronic Revolution

At the end of this introduction, we return to the beginning: the recognition of a technological revolution of unprecedented force unfolding in the context of a global turning point. All dimensions of human existence – the biosphere, technology, and the socio-cultural framework – are exposed to tectonic shifts. The world system is in crisis mode, aptly described as a perma-, omni-, poly-, and planetary crisis. Adding to this crisis mode is the coexistence of numerous systemic and existential risks on a global scale. In this situation, the world system – aptly characterized as a VUCA world – has begun to dangerously vibrate in critical areas and as a whole.

Although each complex system – including the world system – demonstrates a natural resilience to transformative changes, the growing number and intensity of global crisis impulses and their transition into resonance effects, mutual reinforcements, and feedback loops threaten to cross a threshold that could undermine this resilience.

Against this backdrop, one of the great challenges of our time emerges: how can clear problem recognition and solutions be ensured in a world system already operating in crisis mode under such high risk? In fact, there is a manifest chance that, under the ensuing overwhelming pressures, decisions will be made in isolation and under the strain of individual dilemmas into which the broader situation fractures. Compounding this is a tendency to react based on national interests, which are, in turn, dominated by the primacy of security.

So, it comes as no surprise that global solutions are lacking overall when individual risks are downplayed, or their urgency is negated through deliberate strategic decisions. It is to be expected that the regulation of disruptive technologies will often lag behind national competitive and security interests at the respective normative boundaries, more than would be desirable.

However, when dealing with the risks associated with the technetronic revolution, it is not only the demands arising from its life-world context that are challenging but also the inherent impulses of the technetronic revolution itself that can overwhelm societal and professional perception and agency.

Beyond the educational requirements of the technetronic age, what is surely needed is an epistemology that encompasses the significance of this era, recognizing both the significance of the rapid transformations underway and the otherness marking its essence. This should include a heightened awareness of the behavior of complex systems, the historical configuration of the world system and the trajectories of the mega-trends shaping it, and the simultaneity of entire epochs within this historic moment of deep transition.

There is much to suggest that we are riding the crest of a civilizational tsunami, one that – just now – lifts us high enough to glimpse far into the past and deep into the future before the wave crashes over the present.[44] Durante captures the fleeting nature of this moment and one of the great challenges of our time with a quote from Stanley Cavell's *Must We Mean What We Say?* referencing Wittgenstein:

> As I read him, Wittgenstein thought that once a set of practices is ingrained enough to become your form of life, it is difficult to substantively criticise them or even to recognise them as what they are. That's because our form of life is "what has to be accepted, the given." We can no longer get outside of it.[45]

The challenge of this historical moment lies in the fact that the object of our understanding, evaluation, and regulation now exerts such an intense and invasive influence on our capacities to comprehend, act, and respond that it is beginning to reshape the very conditions for these abilities with an almost irresistible force. This prompts the question of how decisively, with what tools, and with which realistic prospects for success we wish to intervene to grasp, adapt to, and meaningfully influence the civilizational change we are witnessing – or whether we should confine ourselves to a fragmented, issue-by-issue approach.

Returning to the ideally framed question of epistemology for the technetronic age, such an approach should encompass an understanding of the deep structural layers and large-scale dynamics of civilization, addressing the transition from one era to the next with an appreciation and deep understanding of the simultaneity of past, present, and future. Reminiscent of the expressive aesthetics of transhumanist tradition, this approach might be described by the term *aeonics*.

NOTES

1 One might also add the progress in fundamental physics and cosmology as intersecting scientific fields expanding to a degree that moves our understanding of reality increasingly beyond comprehension.

2 Similar depictions are those of *homo technologicus, homo biotechnologicus* or *homo deus*. See, for instance, Emil Višňovský, "Homo Biotechnologicus," *Human Affairs* 25, no. 2 (2015): 230–237; Yuval Harari, *Homo Deus: A Brief History of Tomorrow* (New York: Harper, 2017).

3 VUCA World, https://www.vuca-world.org/#:~:text=It%20stands%20for%20Volatility%2C%20Uncertainty,the%20risks%20associated%20with%20them.

4 One might find this foundational distinction of these civilizational dimensions reminiscent of the ancient Greece distinction of *téchnē, phýsis*, and *nomos*.

5 This process is captured by the concept of the Anthropocene as a depiction of the current geological epoch defined by human's destructive shaping of Earth's planetary systems. Paul Crutzen, "Geology of Mankind," *Nature* 415, no. 23 (3 January 2002), http://www.nature.com/nature/journal/v415/n6867/full/415023a.html#top.

6 See for the term "fourth industrial revolution" Klaus Schwab, *The Fourth Industrial Revolution* (New York: Portfolio, 2016).

7 Zbigniew Brzezinski, *Between Two Ages: America's Role in the Technetronic Era* (New York: The Viking Press, 1970), 9.

8 Emil Višňovský, "Homo Biotechnologicus," *Human Affairs* 25, no. 2 (2015): 230.

9 The broad horizon of possibilities from paradise to pandemonium respectively is well-outlined for instance by Max Tegmark, *Life 3.0: Being Human in the Age of Artificial Intelligence* (New York: Knopf, 2017), 161 ff. Overall, the general notion towards the civilizational impact of the technetronic revolution in general and the rapid advancement in computing and AI in particular seems to produce increasingly pessimistic views with some of the major experts changing their initially hopeful assessments into the opposite. This ambivalence is especially evident in the evolving views of AI pioneer Irving John Good. In his seminal 1965 paper, "Speculations Concerning the First Ultraintelligent Machine," Good asserted, "[t]he survival of man depends on the early construction of an ultraintelligent machine." By 1998, however, he had revised this statement, replacing "survival" with "extinction," warning that "because of international competition, we cannot prevent the machines from taking over," as quoted by James Barrat, *Our Final Invention: Artificial Intelligence and the End of the Human Era* (New York: Thomas Dunne Books, 2013), 117. As geopolitical dynamics have shifted from unipolarity to multipolarity in the late 2010s, the urgency and gravity of this concern have only deepened.

10 Peter Bloom, *Identity, Institutions and Governance in an AI World* (New York: Palgrave Macmillan, 2020), 4.

11 Ronald Cole-Turner, "Afterword – Concluding Reflections: Yearning for Enhancement," in *Transhumanism and the Body*, eds. Calvin Mercer and Derek Maher (New York: Palgrave Macmillan, 2014), 175.

12 Ronald Cole-Turner, "Afterword – Concluding Reflections: Yearning for Enhancement," in *Transhumanism and the Body*, eds. Calvin Mercer and Derek Maher (New York: Palgrave Macmillan, 2014), 176.

13 See for the optimistic version, Yuval Harari, *Homo Deus: A Brief History of Tomorrow* (New York: Harper, 2017).

14 Shoshana Zuboff, *The Age of Surveillance Capitalism: The Fight for a Human Future at the New Frontier of Power* (New York: Public Affairs, 2019), 353.

15 Müller refers to nudging, manipulation, and deception as typical techniques of state governmentality which are profoundly questionable from a democratic perspective.

16 See only the milestone publications of the godfather of PR Edward L. Bernays, *Public Relations* (Norman: University of Oklahoma Press, 1952); *Propaganda* (New York: Horace Liveright, 1928); *Crystallazing Public Opinion* (New York: Liveright Publishing Corporation, 1923). In *Public Relations* he summarizes the long history of PR from the "Dark Ages to the Modern World." Edward L. Bernays, *Public Relations* (Norman: University of Oklahoma Press, 1952), 17–128. An interesting theme already in the work of Bernays is also the interchangeability of the consumer and the citizen as target subjects of market research and the public relations based on it.

17 The concept of *panopticon* – originally developed by Jeremy Bentham in the late eighteenth century was originally conceived as a circular prison design where a central watchtower allows a single guard to observe all inmates without them knowing whether they are being watched. This uncertainty leads the inmates to regulate their own behavior, creating a state of self-discipline due to the possibility of constant surveillance. Michel Foucault expanded the idea in *Discipline and Punish: The Birth of the Prison*, trans. Alan Sheridan (New York: Vintage Books, 1977), using the panopticon as a metaphor for modern disciplinary societies and their specific form of "disciplinary power" where surveillance is internalized, leading individuals to self-regulate and behavioral conformity due to the perceived omnipresence of surveillance. See Jeremy Bentham, *The Panopticon Writings*, ed. by Miran Bozovic (London: Verso, 1995); Michel Foucault, *Discipline and Punish: The Birth of the Prison*, trans. Alan Sheridan (New York: Vintage Books, 1977); Ben Godfrey, "The Digital Panopticon: Contemporary

Investigative Methods in Historical Research," in *Investigative Methods: An NCRM Innovation Collection*, eds. Mari Mair, Rachel Meckin, and Martin Elliot (Southampton: National Centre for Research Methods, 2022), 66–74.

18 See for the concept of power-knowledge Michel Foucault, *Discipline and Punish: The Birth of the Prison,* trans. Alan Sheridan (New York: Vintage Books, 1977).

19 See generally for the concept of counter-power Pierre Rosanvallon, *Counter-Democracy: Politics in an Age of Distrust*, trans. Frank Cunningham (Cambridge: Cambridge University Press, 2008).

20 See Wolfgang Fikentscher, *Modes of Thought: A Study in the Anthropology of Law and Religion*, 2nd revised edition (Tübingen: Mohr-Siebeck, 2004), 364 ff. on the cultural dimension.

21 See, for example, the psychological approaches of Melanie Klein, *Envy and Gratitude and Other Works (1946-1963)* (London: Hogarth Press, 1975); Donald W. Winnicott, *Playing and Reality* (London: Tavistock Publications, 1971); Otto F. Kernberg, *Severe Personality Disorders: Psychotherapeutic Strategies* (New Haven: Yale University Press, 1984); or philosophical approaches such as Jean-Paul Sartre, *Being and Nothingness* (New York: Washington Square Press, 1956) (French: L'Être et le Néant, 1943); George Herbert Mead, *Mind, Self, and Society* (Chicago: University of Chicago Press, 1934).

22 See Jean-Paul Sartre, *Being and Nothingness* (New York: Washington Square Press, 1956) (French: L'Être et le Néant, 1943), especially the chapter "The Look" (*Le Regard*), 340 ff.

23 Paolo Prodi, *Settimo non rubare: Furto e mercato nella storia dell'Occidente* (Bologna: Il Mulino, 2009).

24 Michael Lynch, *The Internet of Us: Knowing More and Understanding Less in the Age of Big Data* (London, New York: Liveright Publisher, 2016), 10 (emphasis added).

25 Zuboff argues that the specific governmentality of the technetronic age, which she describes as twenty first-century "surveillance capitalism," distinguishes itself from twentieth-century totalitarianism by the absence of a project to "engineer the soul." Zuboff, *The Age of Surveillance Capitalism: The Fight for a Human Future at the New Frontier of Power* (New York: Public Affairs, 2019), 353. However, despite the lack of a targeted attempt to "engineer our souls," the technetronic revolution and its governmentality could arguably affect our soul in a distinct and lasting way by altering the very conditions under which the fundamental tenets of our collective and individual "soul" are reproduced.

26 Cecilia Åsberg, Redi Koobak, and Ericka Johnson, "Beyond the Humanist Imagination," *NORA – Nordic Journal of Feminist and Gender Research* 19, no. 4 (2011): 218–230.

27 For instance, Carl Gustav Jung, Eric Voegelin, and Hans Jonas have each interpreted modernity in the context of Gnostic thought in their own distinct ways.

28 Other interesting parallels to both Transhumanism and Gnosticism might be explored with regard of alchemistic and occult traditions and thought. Both alchemy and occultism emphasize individual transformation reflecting a deep desire to overcome earthly limitations and explore dualities on esoteric paths. See Stanislas Deprez, "Le transhumanisme est-il une gnose?," *Revue d'éthique et de théologie morale* 302, no. 2 (2019): 29–41; Ethan Richardson, "The New Gnosticism of the Transhumanists," *Mockingbird* (13 June 2017), https://mbird.com/religion/the-new-gnosticism-of-the-transhumanists/; Adam Drakos, "The Gnostic Roots of the Trans Movements," *ThinkingWest* (17 January 2017), https://thinkingwest.com/2022/01/17/gnostic-roots-of-trans-movements/.

29 Gedaliahu G. Stroumsa, "Die Gnosis und die 'christliche Entzauberung der Welt'," in *Max Webers Sicht des antiken Christentums*, ed. Shmuel N. Eisenstadt (Frankfurt am Main: Suhrkamp, 1985), 486–508.

30 As Slavoj Zizek points out in *Living in the End Times*, "life in these apocalyptic times can be characterized by ecological breakdown, the biogenetic reduction of humans to manipulable machines and total digital control over our lives." Slavoj Žižek, *Living in the End Times* (London: Verso, 2011), 327.

31 Again, the sacred view of the body can serve as an interesting test for how transhumanism potentially challenges established religious communities. From the mentioned Taoist perspective on self-cultivation, with the central goal of attaining immortality or at least a state beyond ordinary life, Taoists have been working toward aspirations akin to those of transhumanists for millennia. In Buddhism, the ultimate goal of religious practice is to attain a state beyond the cycles of rebirth, thereby assuming a bodiless and non-physical state. While mere bodily longevity would not align well with Buddhism, it would also not be challenged by transhumanist tendencies toward de-somatization. From a Hindu perspective, merging metaphysics and physics, the flexibility of reincarnation and yogic traditions seeking to actively change the physical condition beyond the normal boundaries of the body make the transhumanist promise seem less alien. Like Christian caution, however, is that of Islam, where the body is a divine creation ultimately owned by its Creator, not to be manipulated in its physical integrity. This view is not far from Judaism, where the body is also considered a holy trust. These very preliminary reflections indicate in line with Geraci and Song that different intellectual contexts respond differently to the technetronic revolution and exhibit varying degrees of compatibility with transhumanist thought. For more, see the contributions in Calvin Mercer and Derek Maher (eds.), *Transhumanism and the Body: The World Religions Speak* (New York, NY: Palgrave Macmillan, 2014).

32 Arguably, the relationship between transhumanism and technological posthumanism is somewhat reminiscent of that between socialism and communism.

33 Other crucial aspects under the broader theme that are not addressed here pertain to the impact of attempt to significantly merge human and machine elements in not-transient ways, the impact of the possibility of machines creating harm to humans, or a possible advent of Singularity with all its consequences for humankind.

34 See for the general topos, Judith Barad, "Blade Runner and Sartre: The Boundaries of Humanity," in *The Philosophy of Neo-Noir*, ed. Mark. T. Conard (Lanham: Lexington Books, 2009), 21–34. Timothy Shanahan, "Blade Runner as Philosophy: What Does It Mean to Be Human?," in *The Palgrave Handbook of Popular Culture as Philosophy*, ed. David Kyle Johnson (London: Palgrave Macmillan, 2020), 983–1003; Philip K. Dick, "Philip K. Dick on Philosophy: A Brief Interview," in *The Shifting Realities of Philip K. Dick: Selected Literary and Philosophical Writings*, ed. Lawrence Sutin (New York: Vintage Books, 1995), 44–47.

35 Shoshana Zuboff, *The Age of Surveillance Capitalism: The Fight for a Human Future at the New Frontier of Power* (New York: Public Affairs, 2019), 351.

36 Adding to the described tendencies are the peculiarities of an increasingly digital economy, where personal contacts and physical locations that traditionally enabled accountability and access within private law practice are diminishing. This shift complicates the mechanisms of interaction and trust, as the digital realm often relies on algorithms and platforms rather than direct human relationships. Consequently, the foundational elements of private law, which depend on personal engagement and tangible connections, face significant challenges in adapting to the evolving landscape of the technetronic age.

37 See Brett Frischmann and Evan Selinger, *Re-Engineering Humanity* (Cambridge: Cambridge University Press, 2018), 61 ff.

38 In this context, she ultimately refers to AI-driven decision systems integrated with sensor-equipped weapons platforms, like drones or armored unmanned vehicles, designed to autonomously patrol, identify, track, and attack targets throughout the entire kill chain.

39 At this level, she highlights the risk of accidental lethal incidents, a danger amplified by the highly systemic and consequential nature of such weapons.

40 Norbert Wiener, *The Human Use of Human Beings* (New York: Free Press, 1989), quoted by Schwarz.

41 An interesting idea worth further exploration is the paradoxical simultaneity of the provisional and inherently error-tolerant notion of the *Hacker Way* with the perfectionist goal of transhumanism.

42 See the dictum "what was once thought can never be unthought," in Friedrich Dürrenmatt's *The Physicists* for a literary exploration of this theme.

43 These considerations often seem absent in the current discourse though. The question of whether we should place an upper limit on robot and artificial intelligence capabilities (see James Barrat, *Our Final Invention: Artificial Intelligence and the End of the Human Era* (New York: Thomas Dunne Books, 2013), 85) appears as futile as the overly optimistic assumption that the reputational risks associated with emulating authoritarian tech practices – such as China's mass surveillance model – would eventually serve as a sufficient deterrent, even for authoritarian regimes. See, however, Steven Feldstein, *The Rise of Digital Repression: How Technology is Reshaping Power, Politics, and Resistance* (New York: Oxford University Press, 2021), 243.

44 For a similar image of a "wave of cultural changes" carrying humanity forward, see Hans Moravec, *Mind Children: The Future of Robot and Human Intelligence* (Cambridge, MA: Harvard University Press, 1988), 9.

45 Stanley Cavell, *Must We Mean What We Say?* (New York: Scribner and Sons, 1969).

REFERENCES

Åsberg, Cecilia, Redi Koobak, and Ericka Johnson, "Beyond the Humanist Imagination," *NORA – Nordic Journal of Feminist and Gender Research* 19, no. 4 (2011): 218–230.

Barad, Judith, "Blade Runner and Sartre: The Boundaries of Humanity," in *The Philosophy of Neo-Noir*, ed. Mark T. Conard (Lanham: Lexington Books, 2009), 21–34.

Barrat, James, *Our Final Invention: Artificial Intelligence and the End of the Human Era* (New York: Thomas Dunne Books, 2013).

Bentham, Jeremy, *The Panopticon Writings*, ed. Miran Bozovic (London: Verso, 1995).

Bernays, Edward L., *Crystallazing Public Opinion* (New York: Liveright Publishing Corporation, 1923).

Bernays, Edward L., *Propaganda* (New York: Horace Liveright, 1928).

Bernays, Edward L., *Public Relations* (Norman: University of Oklahoma Press, 1952).

Bloom, Peter, *Identity, Institutions and Governance in an AI World* (New York: Palgrave Macmillan, 2020).

Brzezinski, Zbigniew, *Between Two Ages: America's Role in the Technetronic Era* (New York: The Viking Press, 1970).

Cavell, Stanley, *Must We Mean What We Say?* (New York: Scribner and Sons, 1969).

Cole-Turner, Ronald, "Afterword – Concluding Reflections: Yearning for Enhancement," in *Transhumanism and the Body*, eds. Calvin Mercer and Derek Maher (New York: Palgrave Macmillan, 2014), 173–191.

Crutzen, Paul, "Geology of Mankind," *Nature* 415, no. 23 (3 January 2002), https://www.nature.com/nature/journal/v415/n6867/full/415023a.html#top.

Deprez, Stanislas, "Le transhumanisme est-il une gnose?," *Revue d'éthique et de théologie morale* 302, no. 2 (2019): 29–41.

Dick, Philip K., "Philip K. Dick on Philosophy: A Brief Interview," in *The Shifting Realities of Philip K. Dick: Selected Literary and Philosophical Writings*, ed. Lawrence Sutin (New York: Vintage Books, 1995), 44–47.

Drakos, Adam, "The Gnostic Roots of the Trans Movements," *ThinkingWest* (17 January 2017), https://thinkingwest.com/2022/01/17/gnostic-roots-of-trans-movements/.

Feldstein, Steven, *The Rise of Digital Repression: How Technology is Reshaping Power, Politics, and Resistance* (New York: Oxford University Press, 2021).

Fikentscher, Wolfgang, *Modes of Thought: A Study in the Anthropology of Law and Religion*, 2nd revised edition (Tübingen: Mohr-Siebeck, 2004).

Foucault, Michel, *Discipline and Punish: The Birth of the Prison*, trans. Alan Sheridan (New York: Vintage Books, 1977).

Frischmann, Brett and Evan Selinger, *Re-Engineering Humanity* (Cambridge: Cambridge University Press, 2018).

Godfrey, Ben, "The Digital Panopticon: Contemporary Investigative Methods in Historical Research," in *Investigative Methods: An NCRM Innovation Collection*, eds. Mari Mair, Rachel Meckin, and Martin Elliot (Southampton: National Centre for Research Methods, 2022), 66–74.

Harari, Yuval, *Homo Deus: A Brief History of Tomorrow* (New York: Harper, 2017).

Kernberg, Otto F., *Severe Personality Disorders: Psychotherapeutic Strategies* (New Haven: Yale University Press, 1984).

Klein, Melanie, *Envy and Gratitude and Other Works (1946-1963)* (London: Hogarth Press, 1975).

Lynch, Michael, *The Internet of Us: Knowing More and Understanding Less in the Age of Big Data* (London, New York: Liveright Publisher, 2016).

Mead, George Herbert, *Mind, Self, and Society* (Chicago: University of Chicago Press, 1934).

Mercer, Calvin, and Derek Maher (eds.), *Transhumanism and the Body: The World Religions Speak* (New York, NY: Palgrave Macmillan, 2014).

Moravec, Hans, *Mind Children: The Future of Robot and Human Intelligence* (Cambridge, MA: Harvard University Press, 1988).

Prodi, Paolo, *Settimo non rubare: Furto e mercato nella storia dell'Occidente* (Bologna: Il Mulino, 2009).

Richardson, Ethan, "The New Gnosticism of the Transhumanists," *Mockingbird* (13 June 2017), https://mbird.com/religion/the-new-gnosticism-of-the-transhumanists/.

Rosanvallon, Pierre, *Counter-Democracy: Politics in an Age of Distrust*, trans. Frank Cunningham (Cambridge: Cambridge University Press, 2008).

Sartre, Jean-Paul, *Being and Nothingness* (New York: Washington Square Press, 1956) (French: *L'Être et le Néant*, 1943).

Schwab, Klaus, *The Fourth Industrial Revolution* (New York: Portfolio, 2016).

Shanahan, Timothy, "Blade Runner as Philosophy: What Does It Mean to Be Human?," in *The Palgrave Handbook of Popular Culture as Philosophy*, ed. David Kyle Johnson (London: Palgrave Macmillan, 2020), 983–1003.

Stroumsa, Gedaliahu G., "Die Gnosis und die 'christliche Entzauberung der Welt'," in *Max Webers Sicht des antiken Christentums*, ed. Shmuel N. Eisenstadt (Frankfurt am Main: Suhrkamp, 1985), 486–508.

Tegmark, Max, *Life 3.0: Being Human in the Age of Artificial Intelligence* (New York: Knopf, 2017).

Višňovský, Emil, "Homo Biotechnologicus," *Human Affairs* 25, no. 2 (2015): 230–237.

VUCA World, https://www.vuca-world.org/#:~:text=It%20stands%20for%20Volatility%2C%20Uncertainty,the%20risks%20associated%20with%20them.

Wiener, Norbert, *The Human Use of Human Beings* (New York: Free Press, 1989).

Winnicott, Donald W., *Playing and Reality* (London: Tavistock Publications, 1971).

Žižek, Slavoj, *Living in the End Times* (London: Verso, 2011).

Zuboff, Shoshana, *The Age of Surveillance Capitalism: The Fight for a Human Future at the New Frontier of Power* (New York: Public Affairs, 2019).

CHAPTER 1

Why Sex Robots Should Fear Us

Nicholas Agar and Pablo García-Barranquero

INTRODUCTION

Should we make sex robots? We explore a line of concern currently underrepresented in this emerging debate. We argue for caution about the development of sex robots out of concern for the harms that may be inflicted on future morally considerable sentient robots. Today's non-sentient sex robots cannot suffer, but they set dangerous precedents for future robots that may be sentient. The harms that future sentient robots could suffer present powerful reasons to refrain from developing sex robots.

We argue that the blurring of boundaries between robots that we are confident lack sentience and robots that could be sentient exacerbates this problem. The sentience of robots is not an unambiguously evident trait. Unlike technological challenges that possess clear criteria for success, such as the quest to send a human astronaut to Mars that can be verified by observation, the construction of a sentient robot lacks clear success criteria. It is the lack of uncontested criteria for sentience that exposes future sentient beings to significant harm. Robert Nozick[1] describes the phenomenon of spillover in which members of one moral category are treated in ways appropriate to another. In the case of sex robots, spillover occurs when we treat robots that could be sentient in ways appropriate to robots that are certainly not sentient.

We do not present these future harms as a decisive moral objection against the creation of sex robots. We note that there are proposals about how to proceed with the technology in a way that might significantly reduce the magnitude of the harm suffered by future sentient robots.[2] Such measures might turn harms that collectively outweigh benefits produced by the creation of sex robots into harms that are tolerable given the benefits that we should expect from them. This paper offers a challenge to advocates of sex robots to present

DOI: 10.1201/9781003226406-2

benefits that both cannot be achieved without the creation of sex robots and are of sufficient moral magnitude to at least compensate for the harms endured by sentient robots.[3]

THE PROBLEM OF ROUGH TREATMENT FOR POSSIBLY SENTIENT ROBOTS

For an indication of where sex robots are in 2020, consider Harmony, a sex robot created by Realbotix – a project dedicated to integrating cutting-edge emerging technologies with silicone doll artistry to provide a bridge between technology and humankind on emotional, mental, and physical levels. Harmony is a fusion of life-like sex dolls and artificial intelligence (AI) that is capable of rudimentary conversation with a bias toward sexual banter.[4]

It is unclear how great the demand is for sex robots. The individuals (typically male) who openly express an interest in acquiring them present them as objects of a scandalized curiosity.[5] One thing we can say is that sex robots are a focus of popular interest in AI and robotics. Sex robots belong to a broader category of social robots, robots that perform social functions by behaving in ways that are similar to humans. It is hard to say how much money there really is to be made from sex robots. But even if these commercial interests are not great, sex robots exercise a disproportionate power over the ways we conceive of artificially intelligent machines. Their influence on the way that we respond to future products of AI is greater than suggested by a count of sex robot sales. Along these lines, Robert Sparrow says: "Sex robots are likely to play an important role in shaping public understandings of sex and of relations between the sexes in the future."[6] Our treatment of future sentient robots could be significantly influenced by our current interest in creating robots to satisfy sexual desire.

The recommendations of this paper occur in the context of a long view of progress in AI. We should view sex robots, among all social robots, as especially susceptible to abusive treatment, yet the harms that we identify occur in the future.[7] They may be experienced by sentient robots we are yet to make.[8] No current sex robot is sentient and it is unlikely that robots we will make over the next decade will be sentient. But machines are acquiring the capacity to perform increasingly many intelligent human behaviors. The existence of future machines capable of the full range of human intelligent behaviors is a reasonable extrapolation given the current progress in AI.[9] Is this sufficient for sentience, though?

There is an additional and distinctively philosophical complexity about whether a machine capable of all intelligent human behavior would be, in actuality, sentient. We grant that the issue of whether a machine that performs all intelligent human behavior is sentient is philosophically controversial and is likely to remain so into the indefinite future. We will refer to these future robots as *possibly sentient* to reflect this philosophical uncertainty. They differ from the certainly non-sentient robots that exist today and those that we expect to produce in the near future. Today's Siri-equipped iPhones are certainly non-sentient. But science fiction offers many examples of robots whose behavior is sufficiently sophisticated for us to entertain the conjecture that they could be sentient.[10] We argue that there is a moral difference between (1) *possibly sentient robots*, those that many experts on AI and the philosophy of mind accept as sentient; and (2) *certainly non-sentient robots*, those that no (or very few) experts on AI and the philosophy of mind accept as sentient.[11]

We should allow that uncertainty about the sentience of future robots exposes robots in the first category to a significant risk of harm. They may face a problem of *rough treatment* due to the history of rough treatment received by their precursor non-sentient sex robots. We define rough treatment as treatment that would cause suffering, were its object sentient. Sex encompasses many acts that threaten rough treatment. Rough treatment directed at certainly non-sentient sex robots may impair their functioning but it cannot cause them to suffer due to their lack of sentience. The problem here is from a failure to distinguish possibly sentient from certainly non-sentient robots.

The problem of rough treatment for possibly sentient robots is well illustrated by this exchange in the Channel 4 fictional TV series *Humans* (2015). Niska is a humanoid robot, a synth. She works as a prostitute and is, unlike the vast majority of synths, sentient. She is questioned about having killed a client. Niska explains that she killed the client because "he wanted to be rough." The lawyer Laura Hawkins responds: "But, is that wrong if he didn't think you could feel?" And she continues: "Isn't it better he exercises his fantasies with you in a brothel rather than take them out on someone who can actually feel?" Niska insists on presenting the client's rough treatment as harmful to her. In what follows, we argue that uncertainty about the suffering conceded by the description *possibly sentient* makes future robots especially susceptible to harm.

REASONABLE EXPECTATIONS OF PROGRESS IN AI

We present the concern about harms inflicted on future sentient robots in the context of the current rapid progress in AI. Realbotix's Harmony is certainly non-sentient. It does not pass the Turing Test.[12] However, it is reasonable to expect that programs capable of realistic human conversation will be loaded into sex robots at some time in the future.

Moravec's Paradox,[13] named after the computer scientist Hans Moravec, expresses a puzzling aspect of advances in AI summarized by Steven Pinker as the idea that, "the hard problems are easy and the easy problems are hard."[14] Computers are now capable of beating the best human players of Go, a strategy game described by its devotees as taking minutes to learn but a lifetime to master. Rodney Brooks notes that these computer-proficient analytic tasks are: "[B]est characterized as the things that highly educated male scientists found challenging, such as chess, symbolic integration, proving mathematical theorems and solving complicated word algebra problems."[15] However, tasks connected with sensorimotor abilities, such as walking through a house, opening doors, and not walking into walls, the kinds of tasks that we typically perform effortlessly, prove very difficult for machines. Even human conversation seems easier for machines to master than human-level sensorimotor behavior. You can ask a Siri-equipped iPhone to identify and offer directions to the nearest restaurant serving Malaysian cuisine.[16] But we are a long way off building a version of Siri that could be fitted to an ambulatory machine directing it to walk to the restaurant, pick up, and expeditiously return your order. This is not to say that human-level sensorimotor capacities will be forever beyond machines, merely that such machines are challenging to make and therefore lie further off in time.

It is likely to be considerably easier to program sex robots to produce convincing human-like verbal behavior than it is to produce the bodily movements of interested human sexual partners.[17] These are things that fit into the "easy for us" but "hard for machines" category of problem. Progress in robotic sensorimotor capacities suggests that these problems will eventually yield to progress in robotics and AI. But we should expect to go through a phase of awkwardly moving robotic sexual partners that are nevertheless capable of engaging in convincing human-like conversations with their purchasers.

Emerging Capabilities of Possibly Sentient Robots

We should distinguish two kinds of uncertainty about the future of AI. There is a distinctively technological variety of uncertainty. The previous paragraphs summarize challenges facing those who seek to make a robot capable of performing the full range of intelligent human behavior. The Turing Test[18] focuses on verbal behavior. Our use of words to express our thoughts makes this focus seem appropriate. This interest leads Turing to exclude evidence about machines that distract from a focus on linguistic behavior. He suggests that we should not be distracted by the fact that computers may be unable "to shine in beauty competitions."[19] The capacity to shine in beauty competitions is more pertinent to designers of sex robots. To be commercially successful, they should also take an interest in the non-verbal behavior of robotic sex partners.

In an echo of the Turing Test, Ben Goertzel, Matt Ikle, and Jared Wigmore propose the "Coffee Test." It addresses the use of intelligence to solve practical challenges. A machine that passes the Coffee Test can "go into an average American house and figure out how to make coffee, including identifying the coffee machine, figuring out what the buttons do, finding the coffee in the cabinet, etc."[20] Machines that pass this practical test of human-level intelligence will respond to the ways in which homes and coffee makers vary with the same facility as humans. There's no reason to think that these practical, sensorimotor abilities will be forever beyond machines. But they do present distinctive challenges. Moravec's Paradox suggests that the Coffee Test will be more difficult to pass than the Turing Test.

We can posit a relative of the Turing Test focused on a robot's capacity to produce the full range of intelligent behavior we would find so important from a human sex partner. We will call this the Masters and Johnson Test for the pioneering sex researchers William H. Masters and Virginia E. Johnson. The Masters and Johnson Test is likely to be significantly harder to pass than the Turing Test. We suspect it will be more difficult to pass than the Coffee Test as well. It will be difficult for a machine to assimilate all of the many ways in which humans make coffee, but human sexual behavior is especially nuanced. A sex robot whose intimate movements are crude approximations to authentic human sexual behavior is likely to be promptly exposed as a fake. In the TV series *Humans*, robots that pass the Masters and Johnson Test coexist with technologies that seem little different from those we have today; Niska, for instance, passes the Masters and Johnson Test. However, this intermingling tends to exaggerate the imminence of possibly sentient sex robots. We will need significant progress in engineering sensorimotor capacities into human sex robots before they could pass the Masters and Johnson Test.

Can Non-conscious Objects Have Sentience?

There is also a distinctive philosophical uncertainty about whether such a robot capable of all intelligent human behavior would be sentient. For the purposes of this paper, we understand sentience as the capacity to feel pain and pleasure. Sentient beings prefer to not experience pain. They prefer to experience pleasure. According to Peter Singer, the philosopher who has done most to make the case for the moral significance of sentience:[21]

> If a being suffers, there can be no moral justification for refusing to take that suffering into consideration (…). If a being is not capable of suffering, or of experiencing enjoyment or happiness, there is nothing to be taken into account. This is why the limit of sentience (…) is the only defensible boundary of concern for the interests of others.[22]

It may be that philosophical obstacles prevent the creation of sentient robots. Consider the qualitative states central to the definition of sentience. Some philosophers have argued that computers may be incapable of qualitative states. Take Jerry Fodor's influential presentation of the computational theory of mind (CTM).[23] Fodor is confident that CTM can explain how a material object could be rational and how it could represent the world. According to Fodor, CTM is silent on what makes a material object conscious. If he is right, then we could be confident that a computer is rational and entertains states with semantic properties without knowing whether it is conscious. Consciousness would be a property of the neurobiological states that fill the computational roles described by the CTM. Thus, a robot that lacks these neurobiological states could be rational and have mental states that represent the world without being conscious. If consciousness is required for sentience then robots whose behavior is directed by a computer, and for that reason lack neurobiological states, will not be sentient. These robots, following Singer's words, will be incapable "of suffering, or of experiencing enjoyment or happiness."[24]

It is appropriate to be epistemically modest about claims about the sentience of sex robots.[25] It is reasonable to grant non-negligible credence to the proposition that we will create robots capable of the full range of human verbal and non-verbal behavior over the next decades. Some philosophers will deny that these robots are sentient. But if they are interested in offering advice about how to treat these beings they should avoid overconfidence about their philosophical conclusions. These skeptics about robot sentience should acknowledge that the arguments of those who defend the possibility of sentient robots are not absurd.[26] Even those whose rejection of the possibility of sentient computers is well-reasoned philosophically should not dismiss the possibility that they might have reasoned in error. It is unlikely that proponents of either side of this debate will achieve the degree of certainty needed to relabel possibly sentient robots as either certainly sentient or certainly non-sentient. This has implications for the advice we give about how to treat future robots that pass the various tests of intelligence. We must approach the possibly sentient robots of the future with the expectation that they could be sentient.

The Ethics of Behavior toward Possibly Sentient Sex Robots

There are circumstances in which this philosophical uncertainty about possibly sentient robots might justify differential treatment. Suppose you confront a dilemma in which you must choose between harming a human or a possibly sentient robot. This kind of practical dilemma features in episodes of *Star Trek: The Next Generation* (1987), in which Data, the possibly sentient android, is expected and expects to place himself in danger to protect human crew members of the *Enterprise.*

Some philosophers will claim that there is no certainty that a psychologically normal human is sentient, especially if we claim that consciousness is constitutive of sentience. Consider, for example, Daniel Dennett's rejection of the existence of conscious feelings.[27] If he is right, then Data's lack of conscious feelings should not give him a moral status inferior to his human crewmates. We should nevertheless consider Dennett's skeptical argument in the broader context of belief about the sentience of machines and humans.

Dennett's arguments may show that we are not justified in certainty about the claim that human brains generate conscious feelings. However, they do not preclude the claim that we are more confident about the conscious feelings of beings with human brains than we are about the conscious feelings of robots directed by digital brains. Dennett's skeptical argument about the existence of conscious feelings applies to both kinds of being. There are skeptical arguments such as the Chinese Room[28] argument of John Searle that find a difference between the sentience of these beings.[29] The existence of these arguments should make us more confident about the sentience of beings with biological brains than we are about beings with digital brains.

Rational doubt about Data's sentience justifies sacrificing him ahead of human crew-members because we are more confident of the sentience of the red-shirted ensign. Data is merely *possibly sentient* while the crewmembers more certainly are. But still, possibly sentient Data does not count for nothing. His crewmembers are right to view harm inflicted on him differently to harm inflicted on certainly non-sentient equipment such as the *Enterprise's* warp drive or teleporter. Perhaps human crewmembers of the *Enterprise* should accept his offers to precede them down dangerous corridors. But they should view him as morally more important than certainly non-sentient machinery. We can think of the rough treatment of Niska in these terms. If, as Laura Hawkins proposes, we were forced to choose between rough treatment of a human and rough treatment of the possibly sentient Niska, we should choose the latter. Of course, it would be better still if neither were treated in this way.

WHY SEX ROBOTS ARE A SPECIAL FOCUS OF MORAL CONCERN?

To summarize up to this point, we can be confident that sentient robots do not exist today. Also, two kinds of uncertainty separate the certainly non-sentient robots of 2020 from the possibly sentient robots of the future. First, technological obstacles must be overcome. Machines that would pass the Masters and Johnson Test for human romantic behavior face sensorimotor challenges more difficult than those involved in passing the Turing Test. We should nevertheless take seriously the possibility that progress in AI will create robots that pass this test for romantic intelligence. Second, there are distinctively philosophical challenges, for instance, uncertainty about whether a sex robot that passes the Masters and Johnson Test is sentient.

Joanna Bryson offers useful advice about how to avoid harming robots by inhibiting their capacity to have sentience.[30] She proposes that we deliberately restrict their psychological development. If we are sure, they cannot suffer then we can be confident that we will not harm them. Bryson's requirement clearly applies to robots we design to work in mines or to defuse bombs. These robots may be expected to perform these complex tasks, but sentience would, at best, be a negligible advantage for extracting ore from a deep mine shaft or in defusing a bomb. Not only would such capacities be surplus to the requirements of their roles but they expose them to risk of harm.

Bryson's quite reasonable advice suggests distinctive problems for all kinds of social robots, including sex robots. Social robots are designed by fill social roles typically filled by humans. A social robot requires some sensitivity to human psychological states. Some social robots may offer company to lonely humans. Others may offer career advice that goes beyond providing statistical breakdowns on the likelihood of success on a given vocational path. Perhaps there will be robotic personal shoppers who accompany shopping expeditions to offer advice on fashion purchases. Humans who make use of social robots value the fact that they resemble them psychologically and physically. This is a feature of sex robots too. Humans who purchase their services want them to be both physically and psychologically like them.

THE DANGEROUS AMBIGUITY OF SEX ROBOT SUFFERING

Sex is a situation in which there is typically an especially strong interest in the mental states of a partner. We have a preference for distinctively human feelings and emotions. For many people, mutual enjoyment is valued part of sex. How often has one human asked a sex partner, "Was that good for you, too?" hoping first that the answer will be "yes!" and more so, hoping that this answer is sincere? This interest is likely to carry over into a future with possibly sentient sex robots. We suspect that Realbotix's certainly non-sentient Harmony is programmed to offer such affirmations when its programming detects that a sexual encounter may have concluded. This interest should lead us to prefer a possibly sentient sex robot's expressions of pleasure to the moans of a certainly non-sentient sex robot.

There is also a dark side to our interest in the mental states of those with whom we engage in sexual acts. Some people enjoy inflicting pain in sex. Sincerity matters here too. Sexual sadists prefer that a partner's cries of pain are authentic. A giggle that follows a cry of pain exposes it as a sham, undermining the principled sadist's enjoyment. A programmed cry of pain coming from a certainly non-sentient sex robot will, for a sincere sadist, be less satisfying that the cry of pain produced by a possibly sentient sex robot.[31]

Vulnerability is an intrinsic feature of robots designed to perform social functions. It is especially valued in sex robots both by humans with conventional preferences and by sexual sadists. Putting to one side sexual exhibitionists, sex is something we typically do in private, rejecting third-party supervision that would intervene should the trust of the possibly sentient sex robot be abused. The privacy of sex provides greater opportunities for abuse, making sex robots more vulnerable to rough treatment than other social robots. Moreover, the fact that rough treatment could cause genuine suffering will predictably appeal to the sexual sadist. We can imagine how an abuser could exploit the philosophical ambiguity of possible sentience. When quizzed about the damage inflicted on the sex robot, rough treatment might be justified by a claim that sentient beings feel nothing. In that case, the screams of a sex robot would not correspond to a negative affective state.

A jury may struggle to find evidence sufficient to convict a client of abuse. Perhaps the sadist will be required to pay to repair damage to a sex robot caused by his or her rough treatment, but a court will be unable to issue a secure judgment that the rough treatment caused suffering. It will be hard to disprove the abuser's claim that roughly treated possibly sentient sex robot feels nothing.[32]

CERTAINLY NON-SENTIENT SEX ROBOTS, POSSIBLY SENTIENT SEX ROBOTS, AND THE PROBLEM OF SPILLOVER

The harms we have identified are in the future. There is no risk of rough treatment harming any current sex robot. iPhones loaded with Siri can be smashed or sexually abused without regard for their moral interests. So, why the interest now in protecting the interests of beings who do not yet exist? In what follows, we offer two reasons that risks imposed on future possibly sentient sex robots should be a concern for us now.

Let us return to our comparison between building a sentient robot and the challenging technological goal of taking human astronauts to Mars. This second goal has the advantage of philosophically clear criteria for success. The safe return of astronauts with a sack full of Martian rock samples is grounds for an unambiguous "mission accomplished." However, this certainly is not the case for the creation of a possibly sentient robot. The goal of making a sentient robot is inherently ambiguous. Few thoughtful observers of progress in AI think that any of its products as of 2020 are sentient. But what of the future? We should expect progress in AI to lead to products whose status as sentient beings is philosophically uncontested. Some appropriately informed judges will assess them as sentient while others will deny this.

The absence of clear, uncontroversial criteria generates a phenomenon Robert Nozick calls spillover.[33] Spillover occurs when we fail to distinguish entities that can be wronged from those that cannot be wronged. Nozick offers his discussion of spillover in the context of the discussion of Kant's view that we should refrain from rough treatment of non-human animals, beings that according to him cannot be harmed, out of concern for beings whom he believes can be harmed. Perhaps torturing animals might lead to character changes making its perpetrators more likely to torture humans. Nozick counters this suggestion with an example involving hitting a baseball. He asks: "If I enjoy hitting a baseball squarely with a bat, does this significantly increase the danger of my doing the same to someone's head?"[34] Spillover would harm people if we made the mistake of treating them as we currently treat baseballs. There are people who bash human heads with baseball bats, but Nozick is confident that this does not result from a failure to appreciate that people and baseballs should receive different treatment. He allows that "it is an empirical question whether spillover does take place or not" but sees no reason why it should "at least among readers of this essay, sophisticated people who are capable of drawing distinctions and differentially acting upon them."[35] Nozick responds to Kant with the suggestion that we can easily distinguish people from animals, and so spillover is unlikely to occur here.

How likely is spillover to occur in the context of sex robots? John Danaher discusses a distinct, especially worrisome kind of spillover generated by today's non-sentient sex robots.[36] He offers the "symbolic-consequences" argument to express this concern. Danaher describes this as the idea that "the robot itself (in its appearance and behavior) and the act of having sex with the robot will have important symbolic properties when it comes to norms of sexual consent and interpersonal sexual ethics."[37] In his presentation, the principal victims of spillover are not future possibly sentient robots but women who find their ability to clearly reject unwanted sexual advances and behavior impaired. The symbolic consequences argument supports the claim that sex robots are "symbolically problematic, both in how they represent human beings and in how they encourage a particular style of sexual interaction with those representations."[38] Danaher suggests that this "problematic symbolism is not essential, or incorrigible, or decisive."[39] He offers suggestions about how we remove and reform it. Danaher suggests that female sex robots do not have to be designed to be "large-breasted, thin-waisted, porn star-esque waifs."[40] Female sex robots need not be programmed to be ever-consenting. They could be programmed "to represent a more progressive set of norms around sexual consent and beauty, and interpersonal relations."[41] Danaher proposes an experiment in which we might try out a number of ways sex robots could be to help determine whether the sex robots can be developed in ways that are overall beneficial.[42] Perhaps some designs of female sex robots could reduce the rate of sexual violence directed at women.

This paper considers a different axis of spillover – harms inflicted on future possibly sentient robots. Steven Spielberg's movie *AI Artificial Intelligence* (2001) offers vivid presentation of spillover. The movie's central character is David, an advanced humanoid robot or mecha, originally adopted by parents as a replacement for a lost child. David seems to qualify as possibly sentient. But other humanoid robots are considerably cruder. Some seem accurately labeled as certainly non-sentient. The movie depicts harms of spillover as the humans express their anger at the technological revolution in AI and robotics by attending "Flesh Fairs" in which they get to cheer on the violent destruction of obsolete mechas. Here rough treatment appropriately directed at certainly non-sentient robots spills over to possibly sentient robots such as David. The distinctions here are considerably more difficult to make than are the distinctions between people and baseballs or between humans and non-human animals.

This argument of this paper allows, as Danaher suggests, that future harms might be prevented or ameliorated. We claim that the harms for future sentient beings could be quite significant. The problem is that possibly sentient robots will be subject to rough treatment that would be entirely appropriate for certainly non-sentient sex robots. One way to prevent harm is to accept the advice of Bryson and ensure prevent robots from developing the capacity for sentience.[43] If this was our collective choice, then we could be confident that rough treatment will not cause future sentient robots to suffer. This avenue of harmful spillover would be blocked. But this seems an unlikely direction for research in AI to take, even if some thinkers think that it would be morally preferable. Our choices about which avenues of robotics to pursue should respond to a possible future in which humans live with possibly sentient robots. The best course of action may be to delay the development of sex robots until we have learned how to treat other social robots with respect.

CONCLUDING REMARKS

This paper presents a moral objection against the creation of sex robots. We have argued that sex robots are a category of social robot that are especially susceptible to rough treatment. Today's certainly non-sentient sex robots cannot suffer. But we should view them as precursors for future possibly sentient robots that can be harmed by rough treatment. Possibly sentient sex robots are threatened by a dangerous ambiguity about their suffering.

NOTES

1 Robert Nozick, *Anarchy, State, and Utopia* (New York: Basic Books, 1974).
2 Neil McArthur, "The Case for Sexbots," in *Robot Sex: Social and Ethical Implications*, ed. John Danaher and Neil McArthur (Cambridge, MA: The MIT Press, 2017), 31–45.
3 See Nicholas Agar, *How to Be Human in the Digital Economy* (Cambridge, MA: The MIT Press, 2019), 110, for an argument that we must respond to the challenge of the digital revolution by working to engineer a matching social revolution. This should respond to the epidemic of social isolation with a matching social revolution that creates jobs that places humans in each other's lives.
4 "My Conversation with Harmony the Sexbot,"August 10, 2017, https://www.cnet.com/videos/realdolls-harmony-ai-sexbot-conversation/, accessed August 7, 2020.
5 Michael Hauskeller, *Mythologies of Transhumanism* (New York: Palgrave MacMillan, 2016), 189.
6 Robert Sparrow, "Robots, Rape, and Representation," *International Journal of Social Robotics* 9, no. 4 (2017): 465.
7 See Nicholas Agar, *The Sceptical Optimist: Why Technology Isn't the Answer to Everything* (Oxford: Oxford University Press, 2015), 10–24, for description of the long view of technological progress that guides this paper's discussion.
8 Michael Hauskeller, *Sex and the Posthuman Condition* (London: Palgrave MacMillan, 2014), 16.
9 David Lawrence, "Robotic Intelligence: Philosophical and Ethical Challenges," in *The Freedom of Scientific Research*, ed. Simona Giordano, John Harris and Lucio Piccirillo (Manchester: Manchester University Press, 2020), 121–132.
10 Isaac Asimov, "Robot Dreams," in *Science Fiction and Philosophy: From Time Travel to Superintelligence*, ed. Susan Schneider (West Sussex: John Wiley & Sons, 2016), 119–124.
11 Sentience is on this view an essentially contested concept. See Walter Gallie, "Essentially Contested Concepts," *Proceedings of the Aristotelian* Society 56, no. 1 (1956): 167–198.
12 The Turing Test is a classical method of determining whether a machine is expected to produce verbal behavior indistinguishable from that of a human. For excellent philosophical presentations of Turing and his test, see Jack Copeland, *Artificial Intelligence: A Philosophical Introduction* (Oxford: Wiley-Blackwell, 1993); Jack Copeland, "The Turing Test," *Minds and Machines* 10, no. 4 (2000): 519–539; Daniel Dennett, "Can Machines Think?," in *How We Know*, ed. Michael Shafto (San Francisco, CA, Harper & Row, 1985), 121–145; Hector Levesque, *Common Sense, the Turing Test, and the Quest for Real AI* (Cambridge, MA: The MIT Press, 2017); James Moor, "An Analysis of the Turing Test," *Philosophical Studies* 30, no. 4 (1976): 249–257; Graham Oppy and David Dowe, "The Turing Test," 2016, https://plato.stanford.edu/archives/spr2019/entries/turing-test/
13 Hans Moravec, *Mind Children: The Future of Robot and Human Intelligence* (Cambridge, MA: Harvard University Press, 1988), 15–16.
14 Steven Pinker, *The Language Instinct: How the Mind Creates Language* (New York: William Morrow & Co, 2007), 190.
15 In Pamela McCorduck, *Machines Who Think: A Personal Inquiry into the History and Prospects of Artificial Intelligence* (Natick, MA: A K Peters, 2004), 456.

16 We also strongly doubt that machines can reflect morally as we do in this paper.

17 See Michael Hauskeller, *Sex and the Posthuman Condition* (London: Palgrave MacMillan, 2014), 11–23.

18 Alan Turing, "Computing Machinery and Intelligence," *Mind* 59, no. 236 (1950): 433–460.

19 Alan Turing, "Computing Machinery and Intelligence," *Mind* 59, no. 236 (1950): 442.

20 Ben Goertzel, Matt Iklé and Jared Wigmore, "The Architecture of Human-like General Intelligence," in *Theoretical Foundations of Artificial General Intelligence*, ed. Pei Wang and Ben Goertzel (Paris: Atlantic Press, 2012), 141.

21 See also Mark Coeckelbergh, "Robot Rights? Towards a Social-relational Justification of Moral Consideration," *Ethics and Information Technology* 12, no. 3 (2010): 210–212.

22 Peter Singer, *Practical Ethics* (Cambridge: Cambridge University Press, 2011), 50.

23 Jerry Fodor, *The Mind Doesn't Work that Way: The Scope and Limits of Computational Psychology* (Cambridge, MA: The MIT Press, 2000).

24 Peter Singer, *Practical Ethics* (Cambridge: Cambridge University Press, 2011), 50.

25 See also Nicholas Agar, "Ray Kurzweil and Uploading: Just Say No!" *Journal of Evolution and Technology* 22, no. 1 (2011): 23–36.

26 See the literature on philosophical disagreement. For example, David Christensen, "Epistemology of Disagreement: The Good News," *The Philosophical Review* 116, no. 2 (2007): 187–217; David Christensen and Jennifer Lackey, *The Epistemology of Disagreement: New Essays* (Oxford: Oxford University Press, 2013); see also Nicholas Agar, "How to Treat Machines that Might have Minds," *Philosophy and Technology* 33 (2019): 1–14, for application of philosophical uncertainty to the problem of whether machines could be sentient.

27 Daniel Dennett, *Consciousness Explained* (Boston, MA: Little, Brown and Company, 1991).

28 Chinese Room is the thought experiment that aims to demonstrate that computers will be forever incapable of thought.

29 John Searle, *Minds, Brains and Science* (Cambridge, MA: Harvard University Press, 1984).

30 Joanna Bryson, "Robots Should be Slaves," in *Close Engagements with Artificial Companions: Key Social, Psychological, Ethical and Design Issues*, ed. Yorick Wilks (Amsterdam: John Benjamins, 2010), 63–74.

31 Ted Chiang, *Exhalation: Stories* (New York: Knopf, 2019).

32 We have some expectation of privacy in a counselling session conducted by a robot-counsellor. But this expectation is weaker than in sex. The news that your counselling session with a possibly-sentient counsellor synth could be subject to supervision by an agency, human or robot, tasked with protecting the welfare of the psychologist is likely to be less off-putting than will the news that your private sexual behavior is subject to monitoring by an external party.

33 Robert Nozick, *Anarchy, State, and Utopia* (New York: Basic Books, 1974), 35–38.

34 Robert Nozick, *Anarchy, State, and Utopia* (New York: Basic Books, 1974), 36.

35 Robert Nozick, *Anarchy, State, and Utopia* (New York: Basic Books, 1974), 36.

36 John Danaher, "The Symbolic-consequences Arguments in the Sex Robot Debate," in *Robot Sex: Social and Ethical Implications*, ed. John Danaher and Neil McArthur (Cambridge, MA: The MIT Press, 2017), 103–112.

37 John Danaher, "The Symbolic-consequences Arguments in the Sex Robot Debate," in *Robot Sex: Social and Ethical Implications*, ed. John Danaher and Neil McArthur (Cambridge, MA: The MIT Press, 2017), 106.

38 John Danaher, "The Symbolic-consequences Arguments in the Sex Robot Debate," in *Robot Sex: Social and Ethical Implications*, ed. John Danaher and Neil McArthur (Cambridge, MA: The MIT Press, 2017), 114.

39 John Danaher, "The Symbolic-consequences Arguments in the Sex Robot Debate," in *Robot Sex: Social and Ethical Implications*, ed. John Danaher and Neil McArthur (Cambridge, MA: The MIT Press, 2017), 114.

40 John Danaher, "The Symbolic-consequences Arguments in the Sex Robot Debate," in *Robot Sex: Social and Ethical Implications*, ed. John Danaher and Neil McArthur (Cambridge, MA: The MIT Press, 2017), 115.

41 John Danaher, "The Symbolic-consequences Arguments in the Sex Robot Debate," in *Robot Sex: Social and Ethical Implications*, ed. John Danaher and Neil McArthur (Cambridge, MA: The MIT Press, 2017), 115.

42 This updates an earlier presentation of the symbolic-consequences argument in which Danaher found its reasoning more persuasive. See, Robert Sparrow, "Robots, Rape, and Representation," *International Journal of Social Robotics* 9, no. 4 (2017): 465, for philosophically useful speculation on the negative consequences for women of robotic representations of rape victims.

43 Joanna Bryson, "Robots Should be Slaves," in *Close Engagements with Artificial Companions: Key Social, Psychological, Ethical and Design Issues*, ed. Yorick Wilks (Amsterdam: John Benjamins, 2010), 63–74.

REFERENCES

Agar, Nicholas, "Ray Kurzweil and Uploading: Just Say No!," *Journal of Evolution and Technology* 22, no. 1 (2011): 23–36.

Agar, Nicholas, *The Sceptical Optimist: Why Technology Isn't the Answer to Everything* (Oxford: Oxford University Press, 2015).

Agar, Nicholas, *How to Be Human in the Digital Economy* (Cambridge, MA: The MIT Press, 2019).

Agar, Nicholas, "How to Treat Machines that Might Have Minds," *Philosophy and Technology* 33 (2019): 1–14.

Asimov, Isaac, "Robot dreams," In *Science Fiction and Philosophy: From Time Travel to Superintelligence*, ed. Susan Schneider (West Sussex: John Wiley & Sons, 2016), 119–124.

Bryson, Joanna, "Robots should be Slaves," in *Close Engagements with Artificial Companions: Key Social, Psychological, Ethical and Design Issues*, ed. Yorick Wilks (Amsterdam: John Benjamins, 2010), 63–74.

Chiang, Ted, *Exhalation: Stories* (New York: Knopf, 2019).

Christensen, David, "Epistemology of Disagreement: The Good News," *The Philosophical Review* 116, no. 2 (2007): 187–217.

Christensen, David, and Jennifer Lackey (eds.), *The Epistemology of Disagreement: New Essays* (Oxford: Oxford University Press, 2013).

Coeckelbergh, Mark, "Robot Rights? Towards a Social-relational Justification of Moral Consideration," *Ethics and Information Technology* 12, no. 3 (2010): 209–221.

Copeland, Jack, *Artificial Intelligence: A Philosophical Introduction* (Oxford: Wiley-Blackwell, 1993).

Copeland, Jack, "The Turing Test," *Minds and Machines* 10, no. 4 (2000): 519–539.

Danaher, John, "The Symbolic-consequences Arguments in the Sex Robot Debate," In *Robot Sex: Social and Ethical Implications*, ed. John Danaher and Neil McArthur (Cambridge, MA: The MIT Press, 2017), 103–112.

Dennett, Daniel, "Can Machines Think?" In *How We Know*, ed. Michael Shafto (San Francisco, CA, Harper & Row, 1985), 121–145.

Dennett, Daniel, *Consciousness Explained* (Boston, MA: Little, Brown and Company, 1991).

Fodor, Jerry, *The Mind Doesn't Work that Way: The Scope and Limits of Computational Psychology* (Cambridge, MA: The MIT Press, 2000).

Gallie, Walter, "Essentially Contested Concepts," *Proceedings of the Aristotelian Society* 56, no. 1 (1956): 167–198.

Goertzel, Ben, Matt Iklé, and Jared Wigmore, "The Architecture of Human-like General Intelligence," In *Theoretical Foundations of Artificial General Intelligence*, ed. Pei Wang and Ben Goertzel (Paris: Atlantic Press, 2012), 123–144.

Hauskeller, Michael, *Sex and the Posthuman Condition* (London: Palgrave MacMillan, 2014).

Hauskeller, Michael, *Mythologies of Transhumanism* (New York: Palgrave MacMillan, 2016).

Lawrence, David, "Robotic Intelligence: Philosophical and Ethical Challenges," In *The Freedom of Scientific Research*, ed. Simona Giordano, John Harris and Lucio Piccirillo (Manchester: Manchester University Press, 2020), 121–132.

Levesque, Hector, *Common Sense, the Turing Test, and the Quest for Real AI* (Cambridge, MA: The MIT Press, 2017).

McArthur, Neil, "The Case for Sexbots," In *Robot Sex: Social and Ethical Implications*, ed. John Danaher and Neil McArthur (Cambridge, MA: The MIT Press, 2017), 31–45.

McCorduck, Pamela, *Machines Who Think: A Personal Inquiry into the History and Prospects of Artificial Intelligence* (Natick, MA: A K Peters, 2004).

Moor, James, "An Analysis of the Turing Test," *Philosophical Studies* 30, no. 4 (1976): 249–257.

Moravec, Hans, *Mind Children: The Future of Robot and Human Intelligence* (Cambridge, MA: Harvard University Press, 1988).

Nozick, Robert, *Anarchy, State, and Utopia* (New York: Basic Books, 1974).

Oppy, Graham, and David Dowe, "The Turing Test," 2016, https://plato.stanford.edu/archives/spr2019/entries/turing-test/.

Pinker, Steven, *The Language Instinct: How the Mind Creates Language* (New York: William Morrow & Co, 2007).

Searle, John, *Minds, Brains and Science* (Cambridge, MA: Harvard University Press, 1984).

Singer, Peter, *Practical Ethics* (Cambridge: Cambridge University Press, 2011).

Sparrow, Robert, "Robots, Rape, and Representation," *International Journal of Social Robotics* 9, no. 4 (2017): 465–477.

Turing, Alan, "Computing Machinery and Intelligence," *Mind* 59, no. 236 (1950): 433–460.

Global Culture for Global Technology

Religious Values and Progress in Artificial Intelligence[1]

Robert M. Geraci and Yong Sup Song

INTRODUCTION

Artificial intelligence (AI) technologies are rapidly becoming pervasive in global culture. From chatbots to surveillance, AI is crucial to big data analysis, corporate communications, government policy, military operations, traffic flow, and even home entertainment. But there is still much room for growth in these technologies' impact on human lives and the trajectory of that impact has yet to be fully determined. We are thus increasingly called upon to reflect on how we design and deploy AI technologies in the interest of human flourishing.

There are clear ways in which the *ideology* of AI is about control and authority, and similarly ways in which specific cultural and religious ideas inform that ideology. The power of AI to control is essential to what Shoshanah Zuboff calls "surveillance capitalism" and also to government regimes directed against both terrorism *and* reasonable citizen dissent.[2] Accomplishing these goals is natural to how AI operates in human networks: perhaps thanks to its origins in cybernetics (the science of "communication and control"), AI clearly fits the model of control expressed in Bentham's panopticon and described by Foucault as a regime of distributed self-regulation.[3] The panopticon was Bentham's design for prisons, schools, hospitals, and other institutions, where the architectural design provides asymmetric information flow between the warden and the inmates (who cannot see when the warden is watching them). The asymmetry of information forces inmates to

DOI: 10.1201/9781003226406-3

act as though they are always under observation, and to thus self-regulate. The potential danger in this has been obvious since the early days of AI, and was considered by early cyberneticists who showed careful attention to the relations of machines to social power.[4]

But in addition to these obvious implications for global power, there are subtler influences on the development of AI. These influences emerge from cultural and religious traditions and can participate in the regimes of surveillance and control. But they might also help mitigate the dangers of authoritarianism, exclusion, economic injustice, and other social evils by pushing AI development along a different trajectory. Different cultures have their own traditions that might be leveraged to bring new and worthwhile values into AI development and deployment. At present, however, the loudest voice belongs to apocalyptic futurists in the United States. While there may be value in their approach to the role of AI in human history, there are other cultures from which we can gain additional sensibilities. After exploring the significance of apocalyptic futurism, this essay notes how specific traditions in China, Thailand, Korea, and India each provide unique contributions to our understanding of AI and thus to our potential future with it. Ultimately, we conclude that a global deployment of AI for human flourishing requires global contributions to the culture and ideology of AI.

APOCALYPTIC AI

The twenty-first century is witness to rising religious or quasi-religious faith in digital salvation thanks to pop science books and the eager advocacy by leading figures in robotics, AI, and digital computing. Drawing on science fiction as well as science, scientists, and technologists like Hans Moravec and Ray Kurzweil produced a movement that anticipates godlike AIs, immortality through mind uploading, and the fulfillment of cosmic purpose. This movement borrows its basic structure and vision of the universe from apocalyptic traditions in Judaism and, especially, Christianity. Thanks to the explosion of Internet access, the pervasiveness of digital computing, and film/television interest in science fiction, the Apocalyptic AI perspective is now the leading ideology of digital futurism.[5] It is so strong that engineers leapt to the idea that Google's LAMDA might be sentient and that other large language models might rapidly become artificial general intelligence.[6]

The Apocalyptic AI[7] worldview borrows from the social and cognitive structures that emerged in ancient Israel under Greek and Roman rule, and which became vital to apocalyptic strains of US Christianity. The ancient texts of 1 Enoch, 2 Baruch, 4 Ezra, and of course Revelation were dominant in this tradition, but the "Markan apocalypse" (Chapter 13 of the Gospel of Mark) and Paul's First Letter to the Corinthians (particularly Chapter 15) were important contributions. The Christian texts – especially Mark 13, 1 Corinthians 15, and Revelation – were decisive in constructing a worldview that largely dissipated from Judaism but which had recurring significance in Christian life. Centuries later in US Christianity, apocalyptic and millenarian beliefs held powerful sway from the arrival of Columbus to the twenty-first century.[8] It is thus no surprise that apocalyptic interpretations of AI emerge in the US context.

The apocalyptic mindset consists of four primary characteristics: (1) a dualistic worldview in which the forces of good battle those of evil but (2) the believer experiences alienation due to the apparent supremacy of evil that will be resolved only in (3) the divine establishment of a glorious world to come and (4) the glorification of the faithful in immortal new bodies to live in the new world.[9] This fourfold structure of Jewish and Christian apocalypticism offered a religious response to the difficulties faced by believers in ancient times (particularly those caused by Greek and Roman rule over Israel), and this same structure has been adopted from Christianity into secular life.

In popular books about robotics and AI, the fourfold apocalyptic structure appears in scientific guise where redemption of humanity and the world is guaranteed by impersonal forces like evolution or the "law of accelerating returns"[10] rather than by divine powers. The progression of apocalyptic thinking follows the same logic but replaces theological concerns over evil and god's redemption with secular existential crisis and scientific solutions. The Apocalyptic AI mindset suggests that human problems, especially mortality, are subject to a digital fix. Paralleling the religious logic of ancient apocalyptic texts, contemporary authors in robotics and AI describe (1) a dualistic struggle between mind and body most apparent in the forms of machine and biology, in which (2) body/biology is presently triumphant over mind as seen by slow learning, poor memory, and the "wanton loss of knowledge"[11] that happens at death, until such time as (3) AI grows to the point that machine intelligence pervades the cosmos accompanied by (4) the technological migration of human minds into machine bodies.[12] Just as the Jewish and Christian apocalypses suggest that history moves inexorably toward the redemption of humanity in a glorious new world, pop science books on robotics and AI promise a digital redemption.

In broad terms, Apocalyptic AI authors propose that advancing technology will create godlike machines and that human beings will join the machine universe as cyborgs or software minds. All authors agree that technological progress has no inherent limits and that machine intelligence will improve until such time as machines equal human intelligence (possibly as early as 2030), after which they will necessarily become far more intelligent than human beings. Supposedly, these machines will be able to simulate entire universes, perform nanoscale interventions (i.e., "miracles") in the physical universe, and spread throughout the cosmos. Apocalyptic AI authors suggest that by upgrading our bodies and brains with cyborg implants, we will "keep pace" with advancing machine intelligence. Further, they believe that through advanced understanding of human bodies and brains, we will fully understand the information pattern that is a human mind – the neurochemical pattern of a human brain – and copy that into immortal machine bodies. Such mind upload will, they argue, provide immortality to individuals. Moravec labels the AI intellects our "Mind Children," whether they emerge entirely from software development or get copied from humanity. He suggests that they will be our evolutionary progeny, the extension of humanity and all of our intellectual virtues into the cosmos. The rise of LLMs, especially the launch of GPT in 2022, accelerated interest and faith in the possibility of fully sentient, greater-than-human artificial intelligence.

The key figures in the rise of Apocalyptic AI include both science fiction (SF) authors and scientists/engineers. Dreams of disembodied intellects go back at least as far as George Bernard Shaw's *Back to Methuselah* cycle of plays (1918–1920) and proceed into Arthur C. Clarke's view of minds downloaded to computer and re-instantiated in cloned bodies in *The*

City and the Stars (1957); such mind transfer would gather steam during the cyberpunk SF of the 1980s and become commonplace by the early twenty-first century. It made memorable appearances in Cory Doctorow's *Down and Out in the Magic Kingdom*, Charles Stross's *Accelerando*, and, perhaps most relevant given the popular 2018 Netflix adaption, *Altered Carbon* by Richard K. Morgan. Scientists' view of the future also drew on dreams of post-human evolution, especially as advocated in various novels by Arthur C. Clarke, including his collaboration with Stanley Kubrick on *2001: A Space Odyssey*. From George Martin's "A Brief Prospect on Immortality" to Hans Moravec's seminal writings in the 1970s, 1980s, and 1990s, such visions of transcendence took on a shiny scientific patina.[13] While a host of other authors played supporting roles in the building of Apocalyptic AI, Moravec contributed to the initial intellectual push which was taken up and popularized by Ray Kurzweil.[14] The currency of this perspective is obvious in subsequent popular discussions of the future of humanity.[15] In 2023, for example, the Future of Life Institute called for a moratorium on large language models and garnered more than 30,000 signatures (as of July 2023), many of which came from eminent researchers in the field.

Apocalyptic AI advocates believe that progress in computing technologies is and will remain exponential, leading to what is called a "Singularity" – a time when technological progress occurs so rapidly that we cannot predict the future of life beyond that point. On our way to that point and beyond, they argue that AI will exceed human capability, even coming to resemble traditional conceptions of the divine, and that human beings will upload their minds into machine bodies and resurrect the dead through computer simulation. Our posthuman descendants (who might be our own digital clones) will allegedly take on the computational prowess of the AIs and join them in a cosmic journey beyond the earth.

ETHICS OF THE END TIMES

The mere possibility of greater-than-human machine intelligence provokes significant debate over the ethics of AI design and the future of human life. Of course, Apocalyptic AI advocates see their project as beneficial to humanity and the cosmos, and thus inherently ethical. For them, solving problems like human mortality and the limits to human intelligence is automatically a morally worthwhile enterprise. Contrary to these promises, however, a growing chorus of leading scientists and technologists suggest that AI could doom humanity.[16] Powerful AI could endanger humanity accidentally, such as by upending political stability or economic systems, or intentionally, should machines decide that humanity poses a danger to themselves. As AI advances and increasingly pervades the global techno-landscape, we must consider how to best intervene in the design process to create trustworthy systems of AI that advance human flourishing.

The concept of ethical AI is multifaceted and contested. At its most basic level, we must differentiate between the ethics of using AI from the ethics built into AI and even from ethics as "learned" or practiced by AI. As an example of the first case, the use of AI for surveillance by police or governments conjures a host of possible concerns from invasion of privacy to prejudicial intervention. Secondarily, machines we build might also provide opportunities to improve upon human decision-making, even in ethical domains, if we successfully program moral values into their operations.[17] Some researchers even believe that AI could promote ethical military engagement.[18] In the final case, if AI takes on human-level or greater intelligence, it will certainly have ethical perspectives of some sort or another. Whether these are consonant with human value systems is open to question. Complicating these thorny questions of AI ethics, different people have different ethical superstructures, assumptions, and ideals based on religion, culture, philosophy, and more. Finding common ground is challenging, and sometimes even impossible. Nevertheless, it is important contemplate how we can attempt such an effort to benefit humanity. Overall, then, advancements in AI raise weighty ethical concerns, even as the ethics themselves are subject to debate.

In *Accelerando*, an influential science fiction book, Charles Stross points toward the significant difficulties we might face in designing ethical AI. In his story, the financial industry and the military are two of the primary drivers of AI advances; and these funding sources are rather obviously ill-suited to building human-friendly, ethical machines. By prioritizing making money or waging war, such systems will naturally leave human beings vulnerable. In *Accelerando*, the machines – known as Vile Offspring (a critical response to Moravec's faith in Mind Children) – threaten the freedom and future of humanity.[19] Stross's Vile Offspring represents the extreme end of AI criticism, as they deliberately set out to enslave and destroy humanity. But the fact that military AI could accidentally cause war, even a global nuclear catastrophe, indicates that the potential for human extinction is built into the way people already choose to employ AI technologies.

Because AI technologies threaten humanity either through accident or intention, a growing chorus of voices publicly advocate for debates over AI ethics. Among the best known of these, Nick Bostrom argues that we must ensure value-alignment between ourselves and machines or else powerful AIs – even if not conscious or generally intelligent – will accidentally threaten us.[20] That is, if AIs do not value things like human dignity, freedom, and life, then they will likely make decisions that preclude those values. This could be malicious or merely accidental, but from our perspective the machines' motive will be irrelevant. In an essay with Eliezer Yudkowsky, Bostrom argues that responsibility, transparency, auditability, incorruptibility, predictability, and easy functionality must all be incorporated into AI.[21] Overall, Bostrom and Yudkowsky note that human intelligence is, even if imperfectly, very significantly general in that it can be applied to a host of unspecified and unanticipated domains. Accomplishing this in a machine is key to establishing safe AI and dictates that advanced, artificial general intelligence be built with the cognition of ethics explicitly incorporated.[22] Given the potential of such AI, Bostrom and Yudkowsky rightly note that we must work toward "an AI which, when it executes, becomes more ethical" than humanity (whatever that might mean), at least through a learning process, even if it is not initially better at moral reasoning.[23]

Religious practitioners also wrestle with the significance of AI, and the values they bring to the conversation have implications for the cultural reception of AI technologies. On the one hand, many Christian believers suggest on the Internet that the pursuit of greater-than-human AI could be reckless and out of sync with Christian values.[24] On the other hand, however, there is a long history of Christians adjusting the domain of salvation to integrate outsiders whom once they excluded and this could plausibly come to include advanced AI.[25] Home devices such as Alexa and Google Home "remember" information gleaned from their owners and can come to imitate the human beings; perhaps by imitating their owners, such devices will take on religious inflections in their communication and expectations about the world.[26] In this fashion, AIs might align with humanity's existing ethical models.

Christian theologians have a more nuanced opinion of Apocalyptic AI than do the online laypeople discussed above, though most Christian theologians also reject the salvation promised in Apocalyptic AI. Noreen Herzfeld, long a leading voice in the intersection of AI and American Christianity, argues that Christian salvation cannot be replaced by cybernetic immortality, but she does so with careful consideration of futurist authors, and without the shrill tone that characterizes many Internet debates.[27] In his introduction to *Transhumanism and Transcendence*, Ronald Cole-Turner lays out the difficult terrain of understanding technological enhancement while contemplating the traditional goal of personal transformation at the heart of Christianity – he does not reject promises of technological transcendence but sees possible modes of consonance.[28] Mirroring this complex connection of Christian theology to transhumanist promises of the future, a growing chorus of voices within Christian theology has emerged. Victoria Lorrimar, for example, sees the apocalyptic visions of AI as a challenge to rethink assumptions and begin a new conversation about novel ways of interpreting Christian perspectives.[29] Michael Morelli notes that AI technology – in his analysis of online chatbots – revels in ambiguity over the future and displaces or translates human intentions.[30]

While many Christian theologians now consider the implications of advanced AI in their theology, some go so far as to radically recast traditional theology with reference to technological promises. Perhaps most futurist among theological anthropologists, Matthew Zaro Fisher suggests – despite his skepticism in the feasibility of mind uploading – that not only could an uploaded personality have a relationship with the Christian God, but the same could be true for AI.[31] Ted Peters, a leading voice in the intersection of theology with science, critiques certain transhumanist promises (e.g., the belief that technical improvements necessarily reshape the moral reality of humankind), but simultaneously acknowledges that technological progress is an important part of humanity's future.[32] For Fisher and Peters, the promises of Apocalyptic AI challenge Christians to a new understanding of their religion. These theologians refuse to see their religious tradition and technological futurism as non-overlapping domains; instead, they suggest that the changes wrought by advancing technology must be part of a divine plan and therefore theology must find ways to account for them. It is even possible that drawing on Christian thought may help guide AI should machines in fact develop greater-than-human intelligence.[33]

The political implications of technological value systems have trouble breaking free from the ethical frameworks of Apocalyptic AI because that ideology runs rampant through US tech culture. The mere possibility of human-equivalent AI conjures significant ethical debate in science fiction, philosophy, and theology. The issues thus raised are magnified if we assume that machines will soon be much smarter than humanity. While specific groups may reflect on AI differently depending on their philosophical, theological, or political perspectives, even a survey as brief as this one points to the vibrant debates in ethics and values that surround AI. While Apocalyptic AI advocates may see a transition from human to machine life as inherently good, others in our society are less sanguine. If nothing else, our efforts to ensure a positive future for humanity demand that we think with care about the values that undergird our technological development.

ALTERNATE EXPERIENCES

So far, this essay has taken its cultural understanding from European and American sources, but the outcomes of global technological development should not be spurred solely by Western social structures.[34] There is no place on Earth where human beings have not struggled to understand the meaning and purpose of human life, and probably no place that lacks an analogous investigation into non-human life (at least in thinking about animals). It is folly to presume that the European and American value-systems offer all of the relevant resources for guiding humanity forward with technologies whose scope is global and whose impact is momentous. We should turn to other cultures and learn what we can from other value systems, which must be rigorously interrogated for their merits even as those in Western contexts must do the same for their own cultural inheritances. While there is insufficient room to do more than hint at the possible contributions of religions outside the fold of Apocalyptic AI, we note that whether we think in terms of traditions (e.g., Buddhism or Taoism) or geography (e.g., Korea or India), there are opportunities to intervene in the ethical debates over AI and supplement what values emerge from Apocalyptic AI.

The local context matters for ideas about technology just as it matters for technologies themselves. Thai philosopher Soraj Hongladarom has pointed toward the importance of local value systems for the deployment of technology, arguing that one nation's values can shape technological development.[35] Hongladarom notes, like Kurzweil, that nanotechnology "is poised to change the very constitution of the body itself" but, rather differently, that "what is needed, in short, is that the goals, agenda and contexts of science and technology should essentially belong to the local culture."[36] Logically, the different religious contexts of the world are not just neutral settings for the installation of advanced AI; they also create opportunities to rethink the deployment of AI globally. Hongladarom is specifically interested in Buddhism, to which we will return shortly, but other religious systems indigenous to Asia also provide insight as we address the growing sophistication of AI.

One approach can be observed in the work of Yueh-Hsuan Weng and his collaborators, who note the potential for a mutually beneficial relationship between religious (as opposed to philosophical) Taoism and AI design. They argue that "AI and robotics could be a feasible way to help Taoist practitioners get closer to their goal of becoming immortals."[37] They articulate this across three domains that clearly emerge out of Apocalyptic AI promises of transcendent AI intelligence. First, they note that post-Singularity AIs will become "akin to gods" and suggest that in such a case it might be possible for Taoists to take advantage of traditional modes of religious interaction to communicate with them.[38] Second, by further extrapolating from Apocalyptic AI, they argue that because religious Taoists aspire toward immortality in their present lives they are uniquely positioned to engage the question of mind uploading.[39] Third, they point toward Taoist models that can help govern AI assistants, provide moral reasoning that connects human beings to such AIs, provide practical modes of instruction, and more.[40] They rightly argue for the importance of religion as a "subfield of culturally-aware robotics" and provide preliminary insight into how religious Taoism might be leveraged in the development of AI.

The Korean concept of *Jeong* – which could be compared to pathos or affect, and is often translated as sympathy, affection, compassion, fellow feeling, solidarity, and forgiveness[41] – is another value that may contribute to the development of AI. *Jeong* originated from a Chinese word and is used in the Korean context under the partial influence of Confucianism. *Jeong* connotes not only the diverse affections toward others, but also the core of Korean social relationships developed in the traditional agricultural society. When *Jeong* emerges in social relationships, it makes them sticky and empathic. For example, Kyu Tae Lee explains *Jeong* by analogy with *meju* beans, which are boiled and fermented soybeans. On the one hand, Lee compared individuals in Western society to raw soybeans, whose relationships are bound by a contract. However, when it is over, they return to each raw soybean. On the other hand, Koreans in traditional agricultural society were compared to *meju* beans. When soybeans are boiled and fermented for a traditional Korean food, *meju,* sticky mucus threads come out of old fermented soybeans and become entangled with each other. *Meju* beans cannot return to their previous state as boiled soybeans. Mucus threads hold the beans together: if one tries to separate the beans, the threads stick together and do not let go easily. Like the sticky mucus of *meju* beans, *Jeong* provides sticky connections over time and space, but also becomes a psychological and emotional link of pain, joy, love and longing that even relates the living and the dead.[42] *Jeong* contributed mainly to strengthening the bonds of the village community.[43] At the same time, however, *Jeong* often expands to nature and to strangers. Finally, *Jeong* defines the scope of the socio-ethical relationships embracing others and, sometimes (without vengeance), even enemies.

For Koreans, *Jeong* represents a key element in human nature. In traditional Korean culture, a person without *Jeong* was often treated as "a human but non-human" [i.e., a human not worthy of being treated as a human being].[44] *Jeong* may contribute to the development of ethical AI as a local, cultural value or a reinterpreted religious value that is new or supplemental to Western Judeo-Christian culture. For example, if the concept of *Jeong* is utilized in the development of AI, it may work as a value that may include AI within human categories of personhood depending on whether we experience reciprocity with AI. In addition, because "many Koreans often feel that *Jeong* is more powerful, lasting, and transformative than love,"[45] *Jeong* may be a strong candidate for a moral value to be utilized for the development of safe AI.

The Indic values of *ahimsa, dharma,* and *swaraj* could be joined to those of other cultures as we promote a just approach to AI. The Indian government has espoused its goals of "AI for All" and "AI for Greater Good" that certainly represent the spirit of technological human flourishing.[46] To go beyond political speechmaking, one must truly engage with the values that get bandied about in broader culture. *Ahimsa* (nonviolence) could be one worthwhile approach to AI, though admittedly it may be too optimistic to believe that governments will concede their "right" to develop autonomous weapons. At a minimum, however, one might demand that the sphere of violence be tightly constrained. For example, economic violence (i.e., the increased marginalization of socio-economically weak classes) could be overcome by public-private partnerships designed around *ahimsa*.[47] Similarly, *dharma* (duty) could be described in terms of mutual obligations and the possibility of technology to provide for one another's needs. Development of AI technologies could be specifically and deliberately aimed toward the satisfaction of such duties. In a final brief possibility, the goal of *swaraj* (self-rule) could reconfigure our approach to AI. Already, some Indian researchers and tech workers hope to build a scientific ethos around "the values of sustainability, plurality and justice" using *swaraj* as their conceptual model.[48] *Swaraj* allows us to pivot our conceptual approach to AI. Present visions of digital empowerment revolve around a person's ability to express him- or herself, to enjoy the capacity to reach others. If we take *swaraj* seriously, along with perhaps using AI to strengthen any given person's reach, we would strengthen each person's control over his or her own life. In this way, empowerment becomes a model of conservation and comfort rather than an opportunity to subject others to one's own whims.[49]

Returning to Hongladarom and the Thai Buddhist context, it could be possible to philosophically reject Western notions of individual selfhood and politically "pay particular attention to the role of compassion, commiseration and on taking concrete action to help others."[50] This appears to place its weight in a very different location compared to the immortalist conjectures of Apocalyptic AI. While advocates of Apocalyptic AI presume the cultural ubiquity of their values, Hongladarom suggests that technologies must account for different values in different contexts. This means that the ideas we have about our technologies are subject to change as they travel the globe, a process that can recursively affect the cultures of origin.

The theological variation in global Buddhism could contribute extensively to our debates over ethics, AI, and technological development. For example, some Buddhists argue that consciousness is not explicitly limited to human beings,[51] and this position could be relevant to legal disputes over the personhood of AIs.[52] Other scholars imply that the goal of *ahimsa* (nonviolence) is somehow integrated into the way that Buddhists (at least online) think about AI and its development[53] and that the Mahayana Buddhist value of compassion could be employed to develop morally superior robots.[54] For the well-known transhumanist thinker James Hughes, there is no cosmic imperative to develop human-equivalent AI, but if we do so we are – from a Buddhist perspective – "obligated to endow them with the capacity for…growth, morality, and self-understanding."[55] The rich history of Theravada, Mahayana, and Vajrayana Buddhist traditions carries a variety of values that could be put in conversation with AI design and the global conversation around AI ethics.

In short, many cultures offer tools that can be used to guide AI development. There is no reason to presume that any one culture has all the necessary tools for thinking about what AI can be and what such technologies can do for humanity. While some wrangling over differing value systems is inevitable, there can and should be opportunities to reflect on our mutual contributions to human flourishing. Here we note that beneficial insights emerge from a variety of geographical and religious contexts; the future of AI and humanity depends on our ability to take advantage of them. The decimation of AI ethics teams in silicon valley, starting with Google firing Timnit Gebru and Margaret Mitchell but extending to such disasters as Microsoft dismantling its top-level AI ethics team in 2023, speaks all the more forcefully to our need for clear discussion of what kinds of regulatory and ethical frameworks should guide ai development from its industrial nucleus to public dissemination.

CONCLUSION

The translation of technologies and technological values across cultures challenges us to work collaboratively in the best interests of humanity. At present, Euro-American cultural values dominate the technological landscape, but these are not the only way to view technologies or to incorporate them into other cultural contexts. Already, there are religious and quasi-religious implications to AI; so, our suggestion that additional values be considered in that domain is not a paradigm shift from secular science to religiously inflected science. Rather, we recognize that science is *already* constituted of religious inflections.[56] In this essay, we suggest that a monolithic approach to the intersections of religion and AI is counterproductive. Rather, we see that the global deployment of AI will benefit from a globally inclusive conversation about the values that guide it.

Obviously, translation from one culture to another is not a new problem. The Internet was early seen as an accelerator for technological deployment and advancement, but it simultaneously required that many countries find internal and external mechanisms for preserving local cultures and identities.[57] Sundar Sarukkai argues that people use intersections with religion to enlarge "the domain of beliefs we hold about technology in order to enable us to deal with it in a manner suited to us."[58] So already there are cultural strategies for domesticating a technology while simultaneously retaining essential practices and beliefs of a culture. In the case of AI, we must find ways to employ the technology in locally meaningful ways but also to guide its usages according to locally and globally beneficial aims.

It could be that an inclusive approach to AI value-systems will promote international cooperation. For example, can nations more firmly entwine their values and technological goals? In doing so, could they find other mechanisms for agreement and collaboration on economic, scientific, and political projects? The transformative outcomes promised by Apocalyptic AI may provide optimism in the face of environmental crisis and political gridlock. To that end, those values may prove vital for the world. But one ought to look also at *swaraj* or compassion as key terms in technological development. Artificial intelligence technologies – from machine learning to big data analysis to commercial robotics, and more – call for careful attention to the values that guide their development. Poor value-alignment will exacerbate existing inequalities or even catastrophically undermine human society. Effective value-alignment requires that we think globally. We must develop AI with an authentically open approach to the world's religious and cultural resources.

NOTES

1 Elements of this chapter were first authored and presented by Robert M. Geraci on October 19, 2019 at the "Governing the Future: Digitalization, Artificial Intelligence, Dataism" conference hosted by the German-Southeast Asian Center of Excellence for Public Policy and Good Governance in Bangkok, Thailand. Prof. Geraci is grateful to Henning Glaser, Dr. Duc Quang Ly, and the CPG staff for the invitation and intellectual collaboration.

2 Shoshana Zuboff, "Big Other: Surveillance Capitalism and the Prospects of an Information Civilization," *Journal of Information Technology* 30, no. 1 (2015): 75–89; Steven Feldstein, *The Global Expansion of AI Surveillance* (Washington, DC: Carnegie Endowment for International Peace, 2019), 11–15. Surveillance capitalism refers to the economic process of using big data acquisition as a form of remuneration for services rendered. For example, Google provides free Internet search and email accounts in exchange for the ability to mine user data and improve its advertising approach.

3 See Robert M. Geraci, *Futures of Artificial Intelligence: Perspectives from India and the U.S.* (New Delhi: Oxford University Press, 2021).

4 Norbert Wiener, *The Human Use of Human Beings* (New York: Houghton Mifflin, 1950); Norbert Wiener, *God & Golem, Inc.: A Comment on Certain Points Where Cybernetics Impinges on Religion* (Cambridge, MA: The MIT Press, 1964; Joseph Weizenbaum, *Computer Power and Human Reason: From Judgment to Calculation* (San Francisco: W.H. Freeman and Company, 1976). Their themes were later taken up by Hayles in her history of cybernetics: N. Katherine Hayles, *How We Became Posthuman* (Chicago: University of Chicago Press, 1999).

5 For example, the Hollywood film *Her* (2013) and popular Netflix shows like *Altered Carbon* and *Black Mirror* describe digital technologies with respect to the kinds of promises made by Kurzweil and his intellectual allies.

6 E.g., Nitasha Tiku, "The Google Engineer Who Thinks the Company's AI Has Come to Life," *Washington Post*, June 11, 2022.

7 In previous publications, one author of this essay explores the intersection of apocalyptic religious perspectives in the development of artificial intelligence and demonstrates the significance of those ideas for policymaking and scientific development. It is therefore necessary only to summarize these findings in this section. See Robert M. Geraci, "Spiritual Robots: Religion and Our Scientific View of the Natural World," *Theology and Science* 4, no. 3 (2006): 229–246; Robert M. Geraci, *Apocalyptic AI: Visions of Heaven in Robotics, Artificial Intelligence and Virtual Reality* (New York: Oxford University Press, 2010); Robert M. Geraci, "Popular Appeal of Apocalyptic AI," *Zygon: Journal of Religion and Science* 45, no. 4 (2010): 1003–1020; Robert M. Geraci, "There and Back Again: Transhumanist Evangelism in Science Fiction and Popular Science," *Implicit Religion* 14, no. 2 (2011): 141–172.

8 For a variety of historical evaluations, see Cathy Albanese, *America: Religions and Religion* (Belmont, CA: Wadsworth, 1999), 168, 477; William D. Apel, 1979. "The Lost World of Billy Graham," *Review of Religious Research* 20, no. 2 (1979): 143; Paul Boyer, *When Time Shall Be No More: Prophecy Belief in Modern American Culture* (Cambridge, MA: Harvard University Press, 1992): 157–162; Rennie B. Schoepflin, "Apocalypticism in an Age of Science," in *The Encyclopedia of Apocalypticism Volume 3: Apocalypticism in the Modern World and the Contemporary Age*, ed. Bernard McGinn, John J. Collins, and Stephen J. Stein (New York: Continuum Press, 2000), 427; and Pauline Moffitt Watts, "Prophecy and Discovery: On the Spiritual Origins of Christopher Columbus's 'Enterprise of the Indies'," *The American Historical Review* 90, no. 1 (1985): 73–102.

9 For a full analysis, see Robert M. Geraci, *Apocalyptic AI: Visions of Heaven in Robotics, Artificial Intelligence and Virtual Reality* (New York: Oxford University Press, 2010), 14–21.

10 Ray Kurzweil, *The Singularity is Near: When Humans Transcend Biology* (New York: Viking, 2005), 7.

11 Hans Moravec, *Mind Children: The Future of Robot and Human Intelligence* (Cambridge, MA: Harvard University Press, 1988), 121.

12 For a full analysis, see Robert M. Geraci, *Apocalyptic AI: Visions of Heaven in Robotics, Artificial Intelligence and Virtual Reality* (New York: Oxford University Press, 2010), 24–36.

13 George Martin, "Brief Proposal on Immortality: An Interim Solution," *Perspectives in Biology and Medicine* 14, no. 2 (1971): 339–340; Hans Moravec, "Today's Computers, Intelligent Machines and Our Future," *Analog* 99, no. 2 ([1976] 1978): 59–84; Hans Moravec, *Mind Children: The Future of Robot and Human Intelligence* (Cambridge, MA: Harvard University Press, 1988), 121; Hans Moravec, *Robot: The Future of Machine and Human Intelligence* (New York: Oxford University Press, 1999).

14 Ray Kurzweil, *The Age of Spiritual Machines: When Computers Exceed Human Intelligence* (New York: Viking, 1999); Ray Kurzweil, *The Singularity is Near: When Humans Transcend Biology* (New York: Viking, 2005), 7.

15 For a variety of later examples, see Max Tegmark, *Life 3.0: Being Human in the Age of Artificial Intelligence* (New York: Knopf, 2017), 161–248; Noah Yuval Harari, *Sapiens: A Brief History of Humankind* (New York: Harper, [2011] 2015), 407–411; Michio Kaku, *The Future of Humanity: Terraforming Mars, Interstellar Travel, Immortality and Our Destiny Beyond Earth* (New Delhi: Allen Lane, 2018), 236–239, 200–205, 218–220; and many of the individual contributions to John Brockman (ed.), *Possible Minds: Twenty-Five Ways of Looking at AI* (New York: Penguin, 2019).

16 A summary of such criticisms can be found at CBInsights, "How AI Will Go Out of Control according to 52 Experts," 2019.

17 For a classic summary of such issues, see Wendell Wallach and Colin Allen, *Moral Machines: Teaching Robots Right from Wrong* (New York: Oxford University Press, 2008).

18 For example, see Ronald Arkin, *Governing Lethal Behavior in Autonomous Robots* (Boca Raton: Chapman and Hall/CRC, 2009).

19 Charles Stross, *Accelerando* (New York: Ace, 2005). Ultimately, humankind finds ways to escape the AIs and provide some hope for a posthuman culture; see Charles Stross, *Accelerando* (New York: Ace, 2005), pp. 380–383, and also Robert M. Geraci, "There and Back Again: Transhumanist Evangelism in Science Fiction and Popular Science," *Implicit Religion* 14, no. 2 (2011): 160–163.

20 Nick Bostrom, "Ethical Issues in Advanced Artificial Intelligence" (2003), https://nick bostrom.com/ethics/ai.html

21 Nick Bostrom and Eliezer Yudkowsky, "Ethics of Artificial Intelligence," in *The Cambridge Handbook of Artificial Intelligence*, ed. Keith Frankish and William M. Ramsey (Cambridge: University of Cambridge Press, 2014), 318.

22 Nick Bostrom and Eliezer Yudkowsky, "Ethics of Artificial Intelligence," in *The Cambridge Handbook of Artificial Intelligence*, ed. Keith Frankish and William M. Ramsey (Cambridge: University of Cambridge Press, 2014), 319–320.

23 Nick Bostrom and Eliezer Yudkowsky, "Ethics of Artificial Intelligence," in *The Cambridge Handbook of Artificial Intelligence*, ed. Keith Frankish and William M. Ramsey (Cambridge: University of Cambridge Press, 2014), 332.

24 Laurence Tamatea, "If Robots R-Us, Who am I: Online 'Christian' Responses to Artificial Intelligence," *Culture and Religion* 9, no. 2 (2008): 141–160.

25 Laura Ammon and Randall Reed, "Is Alexa My Neighbor?" *Journal of Posthuman Studies* 3, no. 2 (2019): 120–140.

26 Laura Ammon and Randall Reed, "Is Alexa My Neighbor?" *Journal of Posthuman Studies* 3, no. 2 (2019): 123.

27 Noreen Herzfeld, *Technology and Religion: Remaining Human in a Co-created World* (West Conshohocken, PA: Templeton, 2009), 64–69, 99–102. See also Noreen Herzfeld, "Cybernetic Immortality versus Christian Resurrection," in *Resurrection: Theological and Scientific Arguments*, ed. Ted Peters, Robert John Russell, and Michael Welker (Grand Rapids, MI: William B. Eerdmans, 2002), 192–201.

28 Ronald Cole-Turner, "Introduction," in *Transhumanism and Transcendence: Christian Hope in an Age of Technological Enhancement*, ed. Ronald Cole-Turner (Washington, DC: Georgetown University Press, 2011), 1–18. The entire volume is composed of essays engaging the relationship between Christian theology and transhumanism, and is thus a notable contribution to the field.

29 Victoria Lorrimar, "Mind Uploading and Embodied Cognition: A Theological Response," *Zygon: Journal of Religion and Science* 54, no. 1 (2018): 191–206.

30 Michael Morelli, "The Athenian Altar and the Amazonian Chatbot: A Pauline Reading of Artificial Intelligence and Apocalyptic Ends," *Zygon: Journal of Religion and Science* 54, no. 1 (2018): 177–190.

31 For Fisher, this means participating in the *imago dei*, or image of God; Matthew Zaro Fisher, "More Human than the Human? Toward a 'Transhumanist' Christian Theological Anthropology," in *Religion and Transhumanism: The Unknown Future of Human Enhancement*, ed. Calvin R. Mercer and Tracy J. Trothen (Santa Barbara, CA: ABC-CLIO, 2015), 23–38. He notes this possibility applies to complex non-human animal species also.

32 Ted Peters, "Progress and Provolution: Will Transhumanism Leave Sin Behind?" in *Transhumanism and Transcendence: Christian Hope in an Age of Technological Enhancement*, ed. Ronald Cole-Turner (Washington, DC: Georgetown University Press, 2011), 63–86. Peters argues that the moral reformation of humanity can only come through a radical break in cosmic history and human life, specifically through divine agency, Ted Peters, "Progress and Provolution: Will Transhumanism Leave Sin Behind?" in *Transhumanism and Transcendence: Christian Hope in an Age of Technological Enhancement*, ed. Ronald Cole-Turner (Washington, DC: Georgetown University Press, 2011), 81–82.

33 Yong Sup Song, "Religious AI as an Option to the Risks of Superintelligence: A Protestant Theological Perspective," *Theology and Science* 19, no. 1 (2020): 65–78.

34 We are alert to the fact that easy demarcations between "western" and "eastern" cultures often obfuscate more than they reveal. For the purposes of this essay, however, we believe the distinction carries enough commonsense meaning to maintain it.

35 Soraj Hongladarom, "Nanotechnology, Development and Buddhist Values," *Nanoethics* 3, no. 2 (2009): 100.

36 Soraj Hongladarom, "Nanotechnology, Development and Buddhist Values," *Nanoethics* 3, no. 2 (2009): 103.

37 Yueh-Hsuan Weng, Yasuhisa Hirata, Osamu Sakura, and Yusuke Sugahara, "The Religious Impacts of Taoism on Ethically Aligned Design in HRI," *International Journal of Social Robotics* 11, no. 5 (2019): 832.

38 Yueh-Hsuan Weng, Yasuhisa Hirata, Osamu Sakura, and Yusuke Sugahara, "The Religious Impacts of Taoism on Ethically Aligned Design in HRI," *International Journal of Social Robotics* 11, no. 5 (2019): 832–833.

39 Yueh-Hsuan Weng, Yasuhisa Hirata, Osamu Sakura, and Yusuke Sugahara, "The Religious Impacts of Taoism on Ethically Aligned Design in HRI," *International Journal of Social Robotics* 11, no. 5 (2019): 833.

40 Yueh-Hsuan Weng, Yasuhisa Hirata, Osamu Sakura, and Yusuke Sugahara, "The Religious Impacts of Taoism on Ethically Aligned Design in HRI," *International Journal of Social Robotics* 11, no. 5 (2019): 834–837.

41 Kyu Tae Lee, *The Psychological Structure of Koreans*. Vol. 2 (Seoul: Shinwon Munwhasa, 1994) [in Korean], 63; Wonhee Anne Joh, "Love's Multiplicity: *Jeong* and Spivak's Notes," in *Planetary Loves: Spivak, Postcoloniality, and Theology*, ed. Stephen D. Moore and Mayra Rivera (New York: Fordham University Press, 2011), 178; Yeol Kyu Kim, "Jeong," *Encyclopedia of Korean Culture* (Seongnam: The Academy of Korean Studies, 1997), http://encykorea.aks.ac.kr/Contents/Item/E0049894 (accessed June 4, 2020) [in Korean].

42 Kyu Tae Lee, *The Psychological Structure of Koreans*. Vol. 2 (Seoul: Shinwon Munwhasa, 1994) [in Korean], 72–74.

43 Yeol Kyu Kim, "Jeong," *Encyclopedia of Korean Culture* (Seongnam: The Academy of Korean Studies, 1997), http://encykorea.aks.ac.kr/Contents/Item/E0049894 (accessed June 4, 2020) [in Korean].

44 Yeol Kyu Kim, "Jeong," *Encyclopedia of Korean Culture* (Seongnam: The Academy of Korean Studies, 1997), http://encykorea.aks.ac.kr/Contents/Item/E0049894 (accessed June 4, 2020) [in Korean].

45 Wonhee Anne Joh, "Love's Multiplicity: *Jeong* and Spivak's Notes," in *Planetary Loves: Spivak, Postcoloniality, and Theology*, ed. Stephen D. Moore and Mayra Rivera (New York: Fordham University Press, 2011), 179.

46 NITI Aayog, *National Strategy for Artificial Intelligence* (Delhi: NITI Aayog, 2018), https://www.niti.gov.in/writereaddata/files/document_publication/NationalStrategy-for-AI-Discussion-Paper.pdf (accessed August 18, 2019). This document does engage the possibility of AI superintelligence (15). The NITI Aayog policy work has been criticized, however, for failing to offer a genuinely inclusive approach to Indian society; see Pakaj Sekhsaria and Naveen Thayyil, "Technology Vision 2035: Visions, Technologies, Democracy and the Citizen of India," *Economic and Political Weekly* 54, no. 34 (2019): 64–69.

47 This approach could also be leveraged in other conversations, such as about the justice system, education, and political representation.

48 *Knowledge Swaraj: An Indian Manifesto on Science and Technology* (Secunderabad: Knowledge in Civil Society Forum, 2011), 5, http://kicsforum.net/kics/kicsmatters/Knowledge-swaraj-an-Indian-S&T-manifesto.pdf.

49 For a further analysis, see Robert M. Geraci, *Futures of Artificial Intelligence: Perspectives from India and the U.S.* (New Delhi: Oxford University Press, 2021).

50 Robert M. Geraci, *Futures of Artificial Intelligence: Perspectives from India and the U.S.* (New Delhi: Oxford University Press, 2021), 104, 105.

51 Laurence Tamatea, "Online Buddhist and Christian Responses to Artificial Intelligence," *Zygon: Journal of Religion and Science* 45, no. 4 (2010): 996. See also Robert M. Geraci, "Spiritual Robots: Religion and Our Scientific View of the Natural World," *Theology and Science* 4, no. 3 (2006): 116–117; Masahiro Mori, *The Buddha in the Robot: A Robot Engineer's Thoughts on Science and Religion*, trans. Charles S. Terry (Tokyo: Kosei, [1981] 1999), 13.

52 Laurence Tamatea, "Online Buddhist and Christian Responses to Artificial Intelligence," *Zygon: Journal of Religion and Science* 45, no. 4 (2010): 992.

53 Laurence Tamatea, "Online Buddhist and Christian Responses to Artificial Intelligence," *Zygon: Journal of Religion and Science* 45, no. 4 (2010): 991.

54 James Hughes, "Compassionate AI and Selfless Robots: A Buddhist Approach," in *Robot Ethics, The Ethical and Social Implications of Robots*, ed. Patrick Lin, Keith Abney, and George A. Bekey (Cambridge, MA: The MIT Press, 2012), 74; Takeshi Kimura, "Masahiro Mori's Buddhist Philosophy of Robot," *Paladyn: Journal of Behavioral Robotics* 9, no. 1 (2018): 74.

55 James Hughes, "Compassionate AI and Selfless Robots: A Buddhist Approach," in *Robot Ethics, The Ethical and Social Implications of Robots*, ed. Patrick Lin, Keith Abney, and George A. Bekey (Cambridge, MA: The MIT Press, 2012), 80.

56 Robert M. Geraci, "A Hydra-logical Approach: Acknowledging Complexity in the Study of Religion, Science, and Technology," *Zygon: Journal of Religion and Science* 55, no. 4 (2020): 948–970.

57 Soraj Honglardarom, "The Web of Time and the Dilemma of Globalization," *The Information Society* 18, no. 4 (2002): 241–249.

58 Sundar Sarukkai, "Culture of Technology and ICTs," in *ICTs and Indian Social Change: Diffusion, Poverty, Governance*, ed. Ashwani Saith, M. Vijayabaskar, and V. Gayathri (New Delhi: Sage, 2008), 47, emphasis removed.

REFERENCES

Albanese, Cathy, *America: Religions and Religion* (Belmont, CA: Wadsworth, 1999).

Ammon, Laura and Randall Reed, "Is Alexa My Neighbor?" *Journal of Posthuman Studies* 3, no. 2 (2019): 120–140.

Apel, William D. 1979, "The Lost World of Billy Graham," *Review of Religious Research* 20, no. 2 (1979): 138–149.

Arkin, Ronald, *Governing Lethal Behavior in Autonomous Robots* (Boca Raton, FL: Chapman and Hall/CRC, 2009).

Bostrom, Nick, "Ethical Issues in Advanced Artificial Intelligence," 2003, https://nickbostrom. com/ethics/ai.html. Revised version of "Ethical Issues in Advanced Artificial Intelligence," In *Cognitive, Emotive and Ethical Aspects of Decision Making in Humans and in Artificial Intelligence Volume 2*, ed. Iva Smit, George Eric Lasker, and Wendell Wallach, 12–17, published by International Institute of Advanced Studies in Systems Research and Cybernetics.

Bostrom, Nick and Eliezer Yudkowsky, "Ethics of Artificial Intelligence," in *The Cambridge Handbook of Artificial Intelligence*, ed. Keith Frankish and William M. Ramsey (Cambridge: University of Cambridge Press, 2014), 316–334.

Boyer, Paul, *When Time Shall Be No More: Prophecy Belief in Modern American Culture* (Cambridge, MA: Harvard University Press, 1992).

Brockman, John (ed.), *Possible Minds: Twenty-Five Ways of Looking at AI* (New York: Penguin, 2019).

CBInsights, "How AI Will Go Out of Control according to 52 Experts," *CBInsights.com* (February 19, 2019), https://www.cbinsights.com/research/ai-threatens-humanity-expert-quotes/ (accessed May 24, 2020).

Clarke, Arthur C., "*The City and the Stars*," in *The City and the Stars and the Sands of Mars* (New York: Warner, [1956] 2001).

Clarke, Arthur C., *2001: A Space Odyssey* (New York: New American Library, 1968).

Cole-Turner, Ronald, "Introduction," in *Transhumanism and Transcendence: Christian Hope in an Age of Technological Enhancement*, ed. R. Cole-Turner (Washington, DC: Georgetown University Press, 2011), 1–18.

Doctorow, Cory, *Down and Out in the Magic Kingdom* (New York: Tor, 2003).

Feldstein, Steven, *The Global Expansion of AI Surveillance* (Washington, DC: Carnegie Endowment for International Peace, 2019).

Fisher, Matthew Zaro, "More Human than the Human? Toward a 'Transhumanist' Christian Theological Anthropology," in *Religion and Transhumanism: The Unknown Future of Human Enhancement*, ed. Calvin R. Mercer and Tracy J. Trothen (Santa Barbara, CA: ABC-CLIO, 2015), 23–38.

Geraci, Robert M., "Spiritual Robots: Religion and Our Scientific View of the Natural World," *Theology and Science* 4, no. 3 (2006): 229–246.

Geraci, Robert M., *Apocalyptic AI: Visions of Heaven in Robotics, Artificial Intelligence and Virtual Reality* (New York: Oxford University Press, 2010).

Geraci, Robert M., "Popular Appeal of Apocalyptic AI," *Zygon: Journal of Religion and Science* 45, no. 4 (2010): 1003–1020.

Geraci, Robert M., "There and Back Again: Transhumanist Evangelism in Science Fiction and Popular Science," *Implicit Religion* 14, no. 2 (2011): 141–172.

Geraci, Robert M., "A Hydra-logical Approach: Acknowledging Complexity in the Study of Religion, Science, and Technology," *Zygon: Journal of Religion and Science* 55, no. 4 (2020): 948–970.

Geraci, Robert M., *Futures of Artificial Intelligence: Perspectives from India and the U.S.* (New Delhi: Oxford University Press, 2022).

Harari, Noah Yuval, *Sapiens: A Brief History of Humankind* (New York: Harper, [2011] 2015).

Hayles, N. Katherine, *How We Became Posthuman* (Chicago: University of Chicago Press, 1999).

Herzfeld, Noreen, "Cybernetic Immortality versus Christian Resurrection," in *Resurrection: Theological and Scientific Arguments*, ed. Ted Peters, Robert John Russell, and Michael Welker (Grand Rapids, MI: William B. Eerdmans, 2002), 192–201.

Herzfeld, Noreen, *Technology and Religion: Remaining Human in a Co-created World* (West Conshohocken, PA: Templeton, 2009).

Honglardarom, Soraj, "The Web of Time and the Dilemma of Globalization," *The Information Society* 18, no. 4 (2002): 241–249.

Honglardarom, Soraj, "Nanotechnology, Development and Buddhist Values," *Nanoethics* 3, no. 2 (2009): 97–107.

Hughes, James, "Compassionate AI and Selfless Robots: A Buddhist Approach," In *Robot Ethics, The Ethical and Social Implications of Robots*, ed. Patrick Lin, Keith Abney, and George A. Bekey (Cambridge, MA: The MIT Press, 2012), 69–84.

Joh, Wonhee Anne, "Love's Multiplicity: *Jeong* and Spivak's Notes," in *Planetary Loves: Spivak, Postcoloniality, and Theology*, ed. Stephen D. Moore and Mayra Rivera (New York: Fordham University Press, 2011), 168–190.

Kaku, Michio, *The Future of Humanity: Terraforming Mars, Interstellar Travel, Immortality and Our Destiny Beyond Earth* (New Delhi: Allen Lane, 2018).

Kim, Yeol Kyu, "Jeong," in *Encyclopedia of Korean Culture*, Seongnam: The Academy of Korean Studies, 1997 [in Korean], https://encykorea.aks.ac.kr/Contents/Item/E0049894 (accessed June 4, 2020).

Kimura, Takeshi, "Masahiro Mori's Buddhist Philosophy of Robot," *Paladyn: Journal of Behavioral Robotics* 9, no. 1 (2018): 72–81.

Knowledge in Civil Society, *Knowledge Swaraj: An Indian Manifesto on Science and Technology* (Secunderabad: Knowledge in Civil Society Forum, 2011), https://kicsforum.net/kics/kicsmatters/Knowledge-swaraj-an-Indian-S&T-manifesto.pdf.

Kurzweil, Ray, *The Age of Spiritual Machines: When Computers Exceed Human Intelligence* (New York: Viking, 1999).

Kurzweil, Ray, *The Singularity is Near: When Humans Transcend Biology* (New York: Viking, 2005).

Lee, Kyu Tae, *The Psychological Structure of Koreans*, Vol. 2 (Seoul: Shinwon Munwhasa, 1994 [in Korean]).

Lorrimar, Victoria, "Mind Uploading and Embodied Cognition: A Theological Response," *Zygon: Journal of Religion and Science* 54, no. 1 (2018): 191–206.

Martin, George, "Brief Proposal on Immortality: An Interim Solution," *Perspectives in Biology and Medicine* 14, no. 2 (1971): 339–340.

Moravec, Hans, "Today's Computers, Intelligent Machines and Our Future," *Analog* 99, no. 2 ([1976] 1978): 59–84, https://frc.ri.cmu.edu/~hpm/project.archive/general.articles/1978/analog.1978.html (accessed March 27, 2019).

Moravec, Hans, *Mind Children: The Future of Robot and Human Intelligence* (Cambridge, MA: Harvard University Press, 1988).

Moravec, Hans, *Robot: The Future of Machine and Human Intelligence* (New York: Oxford University Press, 1999).

Morelli, Michael, "The Athenian Altar and the Amazonian Chatbot: A Pauline Reading of Artificial Intelligence and Apocalyptic Ends," *Zygon: Journal of Religion and Science* 54, no. 1 (2018): 177–190.

Morgan, Robert K., *Altered Carbon* (London: Victor Gollancz, 2002).

Mori, Masahiro, *The Buddha in the Robot: A Robot Engineer's Thoughts on Science and Religion*, trans. Charles S. Terry (Tokyo: Kosei, [1981] 1999).

NITI Aayog, *National Strategy for Artificial Intelligence* (Delhi: NITI Aayog, 2018), https://www.niti.gov.in/writereaddata/files/document_publication/NationalStrategy-for-AI-Discussion-Paper.pdf.

Peters, Ted, "Progress and Provolution: Will Transhumanism Leave Sin Behind?" In *Transhumanism and Transcendence: Christian Hope in an Age of Technological Enhancement*, ed. R. Cole-Turner (Washington, DC: Georgetown University Press, 2011), 63–86.

Sarukkai, Sundar, "Culture of Technology and ICTs," in *ICTs and Indian Social Change: Diffusion, Poverty, Governance*, ed. Ashwani Saith, M. Vijayabaskar, and V. Gayathri (New Delhi: Sage, 2008), 34–58.

Schoepflin, Rennie B., "Apocalypticism in an Age of Science" in *The Encyclopedia of Apocalypticism Volume 3: Apocalypticism in the Modern World and the Contemporary Age*, ed. Bernard McGinn, John J. Collins, and Stephen J. Stein (New York: Continuum Press, 2000), 427–441.

Sekhsaria, Pakaj and Naveen Thayyil, "Technology Vision 2035: Visions, Technologies, Democracy and the Citizen of India," *Economic and Political Weekly* 54, no. 34 (2019): 64–69.

Shaw, George Bernard, *Back to Methuselah (A Metabiological Pentateuch)* (London: Constable, 1921).

Song, Yong Sup, "Religious AI as an Option to the Risks of Superintelligence: A Protestant Theological Perspective," *Theology and Science* 19, no 1 (2020): 65–78.

Stross, Charles, *Accelerando* (New York: Ace, 2005).

Tamatea, Laurence, "If Robots R-Us, Who am I: Online 'Christian' Responses to Artificial Intelligence," *Culture and Religion* 9, no. 2 (2008): 141–60.

Tamatea, Laurence, "Online Buddhist and Christian Responses to Artificial Intelligence," *Zygon: Journal of Religion and Science* 45, no. 4 (2010): 979–1002.

Tegmark, Max, *Life 3.0: Being Human in the Age of Artificial Intelligence* (New York: Knopf, 2017).

Tiku, Nitasha, "The Google Engineer Who Thinks the Company's AI Has Come to Life," *Washington Post*, June 11, 2022.

Wallach, Wendell and Colin Allen, *Moral Machines: Teaching Robots Right from Wrong* (New York: Oxford University Press, 2008).

Watts, Pauline Moffitt, "Prophecy and Discovery: On the Spiritual Origins of Christopher Columbus's 'Enterprise of the Indies,'" *The American Historical Review* 90, no. 1 (1985): 73–102.

Weizenbaum, Joseph, *Computer Power and Human Reason: From Judgment to Calculation* (San Francisco: W.H. Freeman and Company, 1976).

Weng, Yueh-Hsuan, Yasuhisa Hirata, Osamu Sakura, and Yusuke Sugahara, "The Religious Impacts of Taoism on Ethically Aligned Design in HRI," *International Journal of Social Robotics* 11, no. 5 (2019): 829–839.

Wiener, Norbert, *The Human Use of Human Beings* (New York: Houghton Mifflin, 1950).

Wiener, Norbert, *God & Golem, Inc.: A Comment on Certain Points Where Cybernetics Impinges on Religion* (Cambridge, MA: The MIT Press, 1964).

Zuboff, Shoshana, "Big Other: Surveillance Capitalism and the Prospects of an Information Civilization," *Journal of Information Technology* 30, no. 1 (2015): 75–89.

CHAPTER 3

The Utopia of Universal Control

Critical Thoughts on Transhumanism and Technological Posthumanism

Toni Loh

INTRODUCTION

Trans- and posthumanism are two heterogeneous movements of the late twentieth century found in philosophical anthropology and philosophy of technology.[1] At first glance, theorists in these fields, especially transhumanist and technological-posthumanist thinkers, offer very attractive options regarding a potential transformation of the human being into something better: a smarter, happier, more beautiful person – in short, an all-around perfected version of oneself. Promises in this vein, at least implicitly (and far more often explicitly), include assumptions concerning increased autonomy, power, and control over oneself and one's fate. However, I will show in the following that this technological transformation of the human being into a posthuman being is dearly bought, owing to this idea of a universal power of control being accompanied by tendencies of oversimplification, passivation, category error, alienation, and reduction.

To this purpose, I will first differentiate between transhumanism, technological posthumanism, and critical posthumanism, in order to explain my focus on transhumanism afterward and, at last, my focus on technological posthumanism.

Let me summarize my charges in advance. First, oversimplification: transhumanist and (technological-posthumanist) thinking includes the vision of complete power of control over the human via a trivial anthropology and an implicit fundamentalism or technological determinism. Second, passivation: transhumanism (and technological posthumanism) also intends to control the transhumans – those humans that are already partly enhanced, and therefore already in their evolution from the "normal" human being to the posthuman – by including methods of human enhancement as central to its paradigm, thereby passivating the human being as an agent who, during the process of being enhanced, is no longer able to decide on their own development. Third, category error: transhumanist (and technological-posthumanist) thinkers wish to control the ultimate vision of human evolution: the posthuman. In order to reach the goal of concretely describing, understanding, and constructing the posthuman, transhumanism and technological posthumanism are guilty of a category error as they try to describe the posthuman, located in the transcendental realm, in human terms. Fourth, alienation: technological posthumanism, in particular, disapproves of the human body, and in order to completely control the biological substratum of the human, intends to get rid of it as something that "Mother Nature" has saddled us with. Fifth, reduction: technological-posthumanist thinking, in particular, reduces the human mind (and further cognitive competences) to so-called patterns – regular bundles of information that can be completely understood, calculated, controlled, and manipulated – with the goal of uploading them onto computers. This essay intends to give a pointed overview over these five tendencies in order to justify the interpretation of transhumanism and technological posthumanism as utopias of universal control, rather than conducting a detailed analysis that each of the outlined tendencies for themselves would actually require.

AT A GLANCE: TRANSHUMANISM, TECHNOLOGICAL POSTHUMANISM, AND CRITICAL POSTHUMANISM

The transhumanist project is one of developing, enhancing, and perfecting the human being by transforming them. The transhumanist goal is the technological transformation of the human being into a posthuman being, which in the case of transhumanism means a radically modified, "new human being," a human being 2.0, or to be more precise, a human being x.0, since from a transhumanist point of view the potential evolution of the human species is necessarily unfinished. The "trans" in "transhumanism" refers to the attempt to create a better mode of human existence, in working one's way "through" the current human, so to speak.[2] Technics (i.e., technologies and techniques) within the transhumanist paradigm is medium and means for this purpose of optimizing the human being of today to a posthuman being (a human being x.0). Standard transhumanist subjects include, for instance, immortality and radical life extension, as well as methods of human enhancement.

Posthumanism, on the other hand, in general is no longer primarily interested in the human species. Critical posthumanism – as I will call it, following Stefan Herbrechter,[3] Hava Tirosh-Samuelson,[4] and Thomas D. Philbeck[5] – questions the traditional and mostly humanistic dichotomies such as woman–man, nature–culture, and subject–object, that are fundamentally constitutive of our current understanding of the human and the cosmos in general. The critical-posthumanist attempts to go beyond the human by breaking with conventional categories, as well as with their associated vocabulary and thinking. In so doing, critical posthumanism reaches an understanding of the human that is to be located "post" today's essential concept of the human. This, rather than the enhanced human being x.0 of transhumanism, is critical posthumanism's vision of the posthuman.

Between transhumanism and critical posthumanism I'd like to situate a third line of thinking, namely technological posthumanism. Like critical posthumanists, technological posthumanists aren't primarily interested in enhancing the human to a superior version: their vision of the posthuman isn't a radically modified human being – at least not exclusively. But unlike critical-posthumanist thinkers, technological posthumanists don't question humanist categories and dichotomies. On the contrary, they intend to create an artificial alterity, an artificial superintelligence, a strong AI, or universal AI, that in the end will surpass the human species by constituting a new and technological species – this is technological posthumanism's agenda and posthuman vision.[6] On their way to the Singularity – that is, in technological-posthumanist terms, the era that humans will eventually reach by creating an artificial superintelligence – human beings will of course profit from technological achievements, and modify and enhance themselves by means of these advances, for instance, by merging with nanobots, and eventually be immortalized through uploading the human mind onto a computer. But this very transhumanist vision of the transhuman or "transitional human"[7] – those humans who are already in a process of transformation – which serves as a reason for several transhumanists to interpret technological posthumanism as a version of transhumanism, is, within technological-posthumanist thought, merely a nice side-effect and automatic step on the way of the human species toward the posthuman era, rather than the ultimate ambition. Therefore, the role and function of technics within the technological-posthumanist spectrum is to be seen as end, aim and purpose, rather than medium and means (as within transhumanism).[8]

In critical-posthumanist thinking, on the other hand, the technical is neither medium and means nor end, aim and purpose, but rather the principal category (besides culture and the sciences) for criticizing humanist and other traditional categories. Technics within critical-posthumanist thought serves as a category for progressively questioning and merging conventional concepts, thereby crucially influencing and shaping the human beings' understanding of themselves and of the world.

To summarize my thoughts up to this point (see Table 3.1), in trans- and posthumanism there are three strategies to transcend the human being: (1) transhumanism tends to enhance them to a human being x.0; (2) technological posthumanism primarily creates

TABLE 3.1   Trans- and Posthumanism: An Overview.

	Transhumanism	**Technological Posthumanism**	**Critical Posthumanism**
Method	Transformation via technological enhancement	Overcoming via creation	Overcoming via critique
Role / Function of Technics	Medium and means	(primarily) Aim, ends, purpose	Main category of critique
The Posthuman	Human being x.0	(primarily) Artificial alterity	New understanding of the human
Proponents	Nick Bostrom, Stefan Lorenz Sorgner, Max More, James Hughes, Simon Young	Frank Tipler, Marvin Minsky, Ray Kurzweil, Hans Moravec, Vernor Vinge	Katherine Hayles, Rosi Braidotti, Cary Wolfe, Karen Barad, Neil Badmington

an artificial alterity; and (3) critical posthumanism (which I will not talk about in the following) questions the categories that have been conventionally used to define the human.

CRITIQUE OF TRANSHUMANISM

In this section, I will focus on three crucial elements of transhumanist thinking – the transhumanists' understanding of the human, their notion of the transhuman, and their vision of the posthuman – in order to show that transhumanism at its core embeds a utopia of universal control. In this line of thought, transhumanism completely discounts the idea that one essentially cannot control, calculate, and predict human beings. On the one hand, it is true that transhumanism is very much within the horizon of humanism, and numerous famous transhumanists – for instance, Nick Bostrom,[9] Max More,[10] and Simon Young[11] – understand the transhumanist project as a continuation of humanism by technological means. Transhumanism includes several humanist values, such as reason and rationality, freedom and autonomy, the desire for self-design and self-transformation, as well as humanist Enlightenment ideals of Western democratic thinking, such as equality and solidarity, and interprets them at its own discretion. Indeed, Michael Hauskeller claims, "[s]cratch a transhumanist and you will find a humanist underneath."[12]

On the other hand, however, the transhumanist paradigm reveals its most obvious discontinuity with humanist thinking in so far as transhumanism eventually "renounces the Enlightenment's most fundamental premise – that the human being is an end in itself [*Selbstzweck*] and never mere means."[13] Against this backdrop, the critique of transhumanism formulated here is recognizably a humanistic critique. There are other ways of criticizing transhumanist thinking, for instance, in critical-posthumanist terms but in this paper, I restrict my argument to this single form of critique. But to be precise, I am not primarily interested in the notion of control that implicitly pervades transhumanist (and technological-posthumanist) thought, but rather in the idea of a universal power of control being accompanied by tendencies of oversimplification, passivation, and category error.[14]

Controlling the Human: Oversimplification

Transhumanists such as Nick Bostrom[15] and Simon Young[16] appeal to the authority of history to justify their assumption that the human abilities of self-transformation, self-perfection, and self-transcendence are part of the human condition. Bostrom, for example, locates the roots of transhumanist thinking in the Mesopotamian *Epic of Gilgamesh* (2400–1800 BCE), stating that humans naturally attempt to overcome their bodily and mental limitations and restrictions – which in transhumanist thinking are conceived of as intrinsically negative and undesirable.[17] The transhumanist's oversimplified anthropological understanding of human nature and essence is accompanied by an often explicit fatalism regarding the course of history, and especially the course of human evolution.

Though it is true that transhumanists are commonly interested in *individual* enhancement as an expression of personal freedom – sometimes called "freedom of form"[18] or "morphological freedom"[19] – they cannot simply abstract from the *collective* consequences of individual modifications altogether. Every individual choice in terms of self-transformation – even simple enhancements of appearance, style, look, and make-up – shapes society, influences public opinion and educational methods, discriminates, and idealizes. This is far from being trivial, at least when it comes to questions concerning funding, the legislative constitution of human enhancement, and education. These questions and challenges are of course not restricted to the transhumanist paradigm: our humanistic education system, for example (as well as any other education system) sets the framework for what is and what is not to be cultivated regarding the growing human being, and so already gives an answer to these questions. Sadly, only a few transhumanists are genuinely interested in the political and social consequences of their radical enhancement and transformation proposals.[20]

Indeed, even if they suggest that on the collective level things will eventually somehow sort themselves out, transhumanists know that they won't get very far with a purely individualistic focus – at least, not as long as they claim to advocate a philosophy that is not exclusively for a scientific or entrepreneurial elite.[21] Referring to history and human essence is, it seems to me, a purely pragmatic move designed to simplify transhumanist motives and aims for a broader public. Given this aim, transhumanists don't see the need for an anthropologically complex and nuanced understanding of human nature and essence. A simplified anthropology, one restricted to the "instinctive drive"[22] for self-transcendence, is adequate from a transhumanist perspective.[23] And indeed, there exist numerous well-known philosophical approaches that arbitrarily name one attribute to differentiate the "genuine" human from the animal and the divine. The human being has been called the being that can make promises (Nietzsche), the being that acts (Gehlen), the deficient being ("Mängelwesen"; Herder), the *zoon politicon* (Aristotle), and so on. However, these narratives or "essays of definition" ("Definitionsessays")[24] cannot suffice as definitions in a strict sense, and never have; nor do serious philosophical anthropologists justify these narratives as adequate definitions of human nature and essence.

Unfortunately, there are no transhumanist anthropological debates in the current literature, nor are transhumanists interested in answering questions such as "Are those who lack this claimed instinctive drive for self-perfection really human beings?" But the transhumanist trivial anthropology might serve as a reason for the implicit (and occasionally explicit) fatalism that transhumanist thinkers such as Simon Young advocate.[25] The transhumanist line of thinking seems to be that because self-transformation is part of human nature, perhaps even the human *differentia specifica*, we are unable to resist the human being's evolution to a posthuman being. Technological progress is unstoppable, the era of the posthuman "superhuman" is already rising on the horizon of the near future, and those who refuse the transhumanist philosophy are not only theoretically misjudging human nature but practically living in the last deserted "corner" of modernity, together with the last remaining "anti-Facebook primitives" and "anti-Google barbarians," condemned to imminent extinction.

However, the latent fatalism and technological determinism in transhumanist thought is not uncommonly accompanied by an appeal to active participation and deliberate intervention in the "natural evolution" of the human, so as to reach the bright posthuman future even sooner. Transhumanists such as More and Young give their rejection of "Mother Nature's" care a literary form by writing letters to her, and call upon readers to rise up against nature's limitations and restrictions.[26] To control the course of history and the evolution of the human species we first have to understand and control what our human nature basically is. And this is the crux of transhumanism's trivial anthropology.

Controlling the Transhuman: Passivation

Transhumanist thinkers prefer to see their philosophy as a continuation of humanism by technological means.[27] And the most important transhumanist method or category of methods – that is, human enhancement such as physical modifications, mental optimizations, reproductive technologies, genetic enhancement, neuro-technological enhancement, and even moral enhancement[28] – is, according to transhumanists, only a further stage of reaching the humanist ideal of a pedagogical education to empower the human individual to transcend themself. In my view, however, education and cultivation, on the one hand, and human enhancement, on the other, differ in (at least) one crucial aspect.[29]

To quote Thomas Damberger, it is in fact "the primary objective of humanistic enlightenment that the human being will become the root of themself."[30] From this perspective, it is indeed correct to interpret human enhancement methods as a further "expansion of human power to disposition."[31] However, and this is the fundamental difference between humanistic educational methods and strategies of human enhancement, the human being remains at each point in time the agent of humanistic education, whereas transhumanist enhancement degrades the human to the passive material of design and transformation – even in those cases when the to-be-enhanced human being and the enhancing human being are one and the same person. The essential act of enhancing someone uses the to-be-enhanced human being as material. At most, the person in question gives their consent, but in cases such as the reproductive enhancement of an unborn child or the genetic enhancement of future generations, of course none of the parties involved gives their consent beforehand. Here the passivation within the enhancement process is even more obvious.

During their education, humans are at any given moment able to say "No." It is true that an enhanced person can sometimes (depending on the specific circumstances and enhancement method) reverse their enhancement; but the difference between human enhancement and education still holds, in that a reversed enhancement – for instance, a removed implant, a discontinued medication, an aborted treatment – allows the person to return to a relatively similar condition as before the enhancement. No doubt the formerly enhanced person and their body retain the memory of the enhancement, and it would be simply false to suggest that it is the same person as before the enhancement. However, it is much easier to reverse an enhancement than to reverse an educational process. Although people can and do forget, what has once been thoroughly and genuinely learned will tend to stay in a person's memory and remain part of that person from then on.[32]

Regarding human enhancement, the transhumanist desire for control is acute in two respects. On the one hand, the to-be-enhanced human being will be bent to the enhancing human being's will via passivation. That is, humanistic education and human enhancement methods sit at opposite poles of the same axis, in that the more potential there is for autonomy and resistance during the act of enhancement, the closer the practice in question

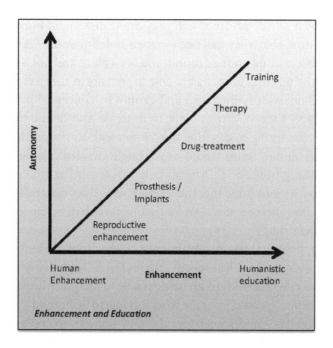

FIGURE 3.1   Trans- And Posthumanism - An Overview

is to conventional humanistic strategies of cultivating the human being (see Figure 3.1). On the other hand, the enhanced human beings retain a personal power of disposal regarding their enhancements, while they lose such power after a completed educational process (this is a matter of degree rather than absolute).

Controlling the Posthuman: Category Error

Numerous transhumanists refer to Giovanni Pico della Mirandola – whose *Oration on the Dignity of Man* (1496), the so-called "Manifesto of the Renaissance," is a central source of humanistic thinking – as an early proponent of transhumanism.[33] This interpretation is in my view correct, in so far as Pico only outlines a *formal* definition of human nature; the human being has, by definition, no place and no fixed rank within the cosmos:

> I have placed you at the very center of the world, so that from that vantage point you may with greater ease glance round about you on all that the world contains. I have made you a creature neither of heaven nor of earth, neither mortal nor immortal, in order that you may, as the free and proud shaper of your own being, fashion yourself in the form you may prefer. It will be in your power to descend to the lower, brutish forms of life; you will be able, through your own decision, to rise again to the superior orders whose life is divine.[34]

The human being is a "chameleon," able to write their own destiny.[35] Human beings can transform into plants and animals as well as into heavenly creatures, and even into the divine, via a four-step evolution from ethics and dialectic through philosophy of nature and metaphysics to theology. The human genuinely has no essence. In giving themself a form and shape by a self-determined act, they are transformed into another being that might be better or worse than they had been in their indefiniteness: "If you see a man dedicated to his stomach, crawling on the ground, you see a plant and not a man."[36] Pico names and defines several beings that humans are able to transform themselves into – except for the divine, which is necessarily ineffable, and cannot be conceived by the human being's intellect and language. Indeed, any attempt to describe and define the divine would be worse than hubris: attempting to describe the divine, with its natural location in the transcendental realm, in human terms would be a straightforward category error. Recall that Immanuel Kant's "thing in itself" ("Ding an sich") faces similar challenges. In the end, it is just an assumption, a pious hope that the human being does not transform into a demon but into the divine, into their creator: "And at last [...] we shall be, no longer ourselves, but the very One who made us."[37]

Transhumanism, however, has no confidence in the transcendental, and attempts to define the posthuman by more or less specific attributes. Admittedly, a few transhumanists, such as Bostrom concede that the human being is able to imagine the posthuman as well as a chimpanzee, for instance, could imagine what it is like to be human.[38] On the other hand, countless transhumanists – and sometimes the very same thinkers[39] – claim the posthuman to be equipped with specific capacities and abilities that are qualitatively gradable (compared to the capacities and abilities of the human 1.0) until at some (unknown) point in human evolution the human species will vault the categorical abyss between the human and the posthuman (located, according to Pico della Mirandola, between the heavenly creature and the divine). As Bostrom puts it

> I shall define *a posthuman* as a being that has at least one posthuman capacity. By *a posthuman capacity*, I mean a general central capacity greatly exceeding the maximum attainable by any current human being without recourse to new technological means. I will use *general central capacity* to refer to the following: *healthspan* [...], *cognition* [...], *emotion* [...].[40]

On the other side of that abyss, the transformation – or, to be more precise, the posthumanization – of the posthuman seems to be thought of as continuously proceeding.[41] Transhumanists have lost Pico's and Christianity's faith in not only the intangible but – and especially – the wonder and glory of the divine. They commit a category error because of their efforts to control not only the development of the human and transhuman, but also the posthuman. Here it becomes unclear to what extent the transhumanists' vision of the posthuman is just a "new human being" and to what extent the posthuman is a human no longer.[42]

CRITIQUE OF TECHNOLOGICAL POSTHUMANISM

In this section I focus on technological posthumanism, which can be situated midway between transhumanism and critical posthumanism owing to its similarity to transhumanism in substance and content, and its kinship with critical posthumanism regarding the transcending of the human. There are numerous family resemblances between transhumanism and technological posthumanism, so many in fact that technological posthumanism gives the impression of being a loose and open bundle of ideas revolving around the posthuman vision of an artificial superintelligence rather than a theory capable of sustaining an autonomous line of thinking. Systematically and methodologically, technological posthumanism is built on "sandy ground," and the many noticeable borrowings of theoretical fragments from transhumanism cannot completely mask a lack of theoretical complexity. In light of this, several thinkers – especially transhumanists such as Stefan Lorenz Sorgner, Max More and Nick Bostrom – understand technological posthumanism as an extreme version of transhumanism in some respects.[43] In fact, technological posthumanists could be seen as even "better" transhumanists in that they extend the transhumanist drive for an ultimate power of control to further aspects of human existence. For example, the technological-posthumanist Ray Kurzweil promises humans, on entering the Singularity, complete "power over [their] fates."[44]

My critique of technological posthumanism is for these reasons completely in accord with my reflections regarding transhumanism and its notion of control via oversimplification, passivation, and category error. In the following, I will concentrate on two aspects of technological-posthumanist thinking that don't play such an essential role within the transhumanist paradigm: technological posthumanists' negative attitude toward the biological body and their reduction of human nature and mind to information. Both of these elements illustrate that technological posthumanism is in fact very much in line with the transhumanist goal of universal control. However, within transhumanist thinking, the human body can become transformed into a posthuman – the human being x.0 of the posthuman era and condition is imagined by transhumanists to be equipped with a body. Here, transhumanism is much closer to humanism than technological posthumanism is. Moreover, many transhumanists as well as technological posthumanists favor of a view of the human mind and personality called "patternism"[45] – an approach that understands all non-material human competences such as mind, soul, reason, personality, and the like, which are commonly located in the human brain, to be independent of any substance and hence transferable to any medium. But, unlike technological posthumanists, transhumanists still tend to be more open toward other interpretations of the human mind and personality.

Controlling the Human Body: Alienation

A crucial element, and related field of research, of the technological-posthumanist paradigm concentrates on technological posthumanists' hope that one day humans will be able to upload their mind (soul, personality, etc.) onto computers. Approaches going under the name of Mind Uploading, Mind Downloading, Mind Copying, Mind Cloning, Mind Transfer, Whole Brain Emulation, and Whole Brain Simulation share the premise that the biological body is not important for human beings.[46] Some technological posthumanists even go so far as to call the human body a "piece of meat," "pulp," or "jelly." To put it briefly, the human biological body does not have a good reputation within technological-posthumanist thought. As a first step in the transformation process, it is of course necessary to optimize, modify and radically enhance the biological substratum of the human – in typical transhumanist manner – to override the bodily conditions and to equip them with new abilities and competences. However, in finally entering the Singularity, the human mind (soul, personality, etc.) will completely abandon its frail and mortal shell to move into an artificial home that the person in question has shaped and designed beforehand. In Hans Moravec's words, the "postbiological"[47] human will have a silicon-based body, whereas the genuine posthuman of an advanced singular era will no longer need any material basis whatsoever.[48] The pure mind is at this stage free to merge with other minds, with any matter (now animated, according to Moravec),[49] and ultimately with the awakened universe (following Kurzweil).[50]

The desire to be able to abandon the biological body, and finally any body at all, is much stronger in technological posthumanism than it is in transhumanism. In transhumanist thinking, the human body is still part of human self-understanding. Technological posthumanism, on the other hand, ultimately reduces the human to their mind (soul, personality, etc.), which is still human (although postbiological) after it has left its biological "grave" and abandoned any bodily shell whatsoever. The mind-body dualism implicitly pervading transhumanism is biased in favor of the human mind, although the human body is not completely absent from the reckoning in the transformation to a posthuman being. Nevertheless, technological posthumanists fully express their mind-body dualism in support of the human mind.

This somatophobic attitude of technological posthumanism is, in addition to the examples given in the last section, yet another example of its overall pursuit of control: what one has created oneself is completely transparent and can therefore be wholly governed by powers at the creator's disposal. Organic nature, on the other hand, still has the "hem of its skirt" in the "mystical limbo" of growth and decay; these are processes that we will never fully understand and, therefore, will never be able to control completely. From this unsophisticated anti-phenomenological position, technological posthumanists are (from the perspective) above such "messy" concepts as the "Leib" of Edmund Husserl and Maurice Merleau-Ponty, which attempts to bridge the abyss between mind and body. They really believe that what they define as mind is wholly able to exist independently from the body in the very same way it does in conjunction with the body – and even better: the pure mind will be able to realize its full potential without bodily bonds. This latter thought is precisely Ray Kurzweil's reason for calling superintelligent machines and "nonbiological forms of intelligence" even "more exemplary of what we regard as human than it is today."[51] Nonbiological forms of intelligence are purer, clearer, and stronger – thanks to their independence from every biological basis – than the human mind, which is still stuck in "Mother Nature's" organic "swamp." Artificial superintelligences could already realize human potential. If humans follow the machines on this path into a postbiological mode of existence, they would in fact humanize themselves, realize their actual human potential, *become* human, so to speak, before saying farewell to every silicon-based body, to any body whatsoever, and eventually leaping over the abyss toward the posthuman.

This picture of a deficient body that is not essentially human, but rather something that "Mother Nature" has contingently saddled us with, is the root of technological posthumanists' urge toward an overall power of control. In transhumanism the "whole" human being (i.e., his mind and body) is deficient, whereas in technological posthumanism it is not really the human that is deficient, because human essence is genuinely their mind. And this mind could have managed its entrance into Singularity long ago, had it not been confined to this mortal shell of a body. Hence, according to technological posthumanism it is necessary to get control over this alien body that is not us, to master it, to shape it, and eventually to abandon it.

Incoherently, technological-posthumanist thinkers implicitly transfer characteristics of the biological substratum of the human (such as the five senses) to the silicon-based body, as well as to their understanding of the virtual sphere in imagining the human mind (soul, personality, etc.) as an entity freed from any bodily limitations and restrictions.

Controlling the Human Mind: Reduction

A common approach within technological-posthumanist theory is that of "patternism" (due to Susan Schneider and others), a view that understands all non-material human competences such as mind, soul, reason, personality, and the like, which are ordinarily located in the human brain, to be independent of any substance, and because of that fact transferable to any medium. Technological posthumanists orient themselves with this approach to John Locke's understanding of personal identity as basically consisting of memories and to Thomas Hobbes's definition of "reason" as "reckoning." Consequently, the technological-posthumanist concludes that a person's identity and mind consist of small, clearly knowable, and therefore calculable units – patterns, information – that are independent of any material basis.

The technological-posthumanist project of mind uploading and, via this procedure, completely computerizing one's personality (thereby creating a detailed copy of oneself) satisfies the desire for control of the human body, as already explained above. But further, technological posthumanists believe that they will eventually be able to fully govern everything non-material. The transhumanist Martine Rothblatt illustrates this perspective by claiming that consciousness could be artificially produced by translating individual attributes such as mannerisms, personal characteristics, feelings, and values into so-called "bemes."[52] Such thinkers invite the human mind not only to enter the virtual "wonderland" via information channels, but to travel in space, to wander between the stars and the planets.[53] Information as the ontological "lowest common denominator" of reality is not only the essential substance of the human mind, but also of any non-material entity in the cosmos. Hence, even animals could upload themselves, travel between the planets, and maybe even enter the Sinuglarity.[54] Although some transhumanists invite non-human creatures to the posthuman condition, transhumanism is constrained by an anthropocentrism inherited from humanism.[55] In contrast, technological posthumanism – which is no longer primarily interested in human beings – smooths the way for animals and plants to enter the Singularity – all for the price of a radical mind-body dualism, or to be more precise, informational monism and a reductionist understanding of any non-material entity.

As already sketched with the help of Martine Rothblatt's approach, this definition of personal identity and mind as consisting of calculable and predictable information units allows for the complete control of everything that technological posthumanists understand to be essentially human. Hannah Arendt not only criticizes this Hobbesian definition of reason as reckoning in the natural sciences as a dangerous tendency to oversimplify, to replace logical critical reflection and judgment with logical calculation, to deny responsibility for the public sphere and the human world;[56] she also criticizes the approach of "patternism" (although of course not literally) in her *Vita activa*, differentiating between the Who and the What of a person. According to Arendt, we are never able to understand and to talk about the Who of a person, only about the What, for instance by listing their physical and mental attributes, likes and dislikes, and idiosyncrasies. The Who of a person, on the other hand, is simply there; we are confronted with it, it unsettles us, and it is necessarily beyond any "control."[57] Whenever we open our mouth to talk about someone, to describe them to another person, to explain why we love or hate them, and so on, it is exclusively the What we are able to put into words. Technological-posthumanist thinkers try to get hold of this unavailability of the human person and mind by translating it into information. However, in listing various human qualities, people such as Rothblatt, Moravec, and Minsky are only able to control the human What. The technological-posthumanist dismemberment of the human into a conglomeration of attributes, which could, like a puzzle, be uploaded via a computer interface, sent to another planet, and be put back together again to recreate the very same person, reduces the human being to Arendt's What. But, even if this procedure could work, this artificial copy – or whatever it might be that emerges in our new virtual sphere of existence – will, of course, have a Who as well. This Who can, I believe, never be fully controlled.

CONCLUSION

To summarize my discussion in this essay: At first, in this analysis of transhumanism and technological posthumanism, I differentiated between transhumanism, technological posthumanism, and critical posthumanism. After that, I focused on three crucial elements of transhumanist thinking – the transhumanists' understanding of the human, their notion of the transhuman, and their vision of the posthuman – in order to show that transhumanism at its core embeds a utopia of universal control. Clearly, the transhumanist paradigm engenders tendencies of oversimplification, passivation, and category error. At last, I extended this interpretation to the spectrum of technological posthumanism, which is – at least in some respects – very similar to transhumanist thinking, in order to show that technological posthumanists are even more strongly driven by this impulse toward control than transhumanists are. Technological posthumanists' attempts to control the human body show clear tendencies of alienation, while their attempts to control the mind through the techniques of "patternism" and "mind uploading" reveal their predilection for simplistic reduction.

For anyone thinking of answering the transhumanists' and technological posthumanists' call to transform into a better version of themself and eventually enter the Singularity, the price they must pay is an overly simplistic definition of human nature, the passivation into bare material while being enhanced, a category error in attempting to describe the posthuman being, alienation from one's body, and a reduced understanding of one's mind.

NOTES

1 Bernhard Irrgang, *Posthumanes Menschsein? Künstliche Intelligenz, Cyberspace, Roboter, Cyborgs und Designer-Menschen – Anthropologie des künstlichen Menschen im 21. Jahrhundert* (Stuttgart: Franz Steiner Verlag, 2005); Janina Loh, *Trans- und Posthumanismus zur Einführung* (Hamburg: Junius, 2018).

2 Dieter Birnbacher, "Transhumanismus – Provokation oder Trivialität?" in Stefan Lorenz Sorgner (ed.), *Aufklärung und Kritik: Zeitschrift für freies Denken und humanistische Philosophie: Schwerpunkt Transhumanismus* 22, no. 3 (2015): 130–151; Dieter Birnbacher, "Posthumanity, Transhumanism, and Human Nature," in *Medical Enhancement and Posthumanity*, ed. Bert Gordijn and Ruth Chadwick (Berlin and New York: Springer, 2008), 95–106; Joel Garreau, *Radical Evolution: The Promise and Peril of Enhancing our Minds, Our Bodies – And What It Means to be Human* (New York: Broadway Books, 2005), 231–232; Christian Niemeyer, "Vom Transhumanismus zurück zur Transformation, altdeutsch: 'Verwandlung'," in Stefan Lorenz Sorgner (ed.), *Aufklärung und Kritik: Zeitschrift für freies Denken und humanistische Philosophie: Schwerpunkt Transhumanismus* 22, no. 3 (2015): 130–144; Thomas D. Philbeck, "Ontology," in *Post- and transhumanism: an introduction*, ed. Robert Ranisch and Stefan Lorenz Sorgner (Frankfurt am Main: Peter Lang, 2014), 173–183; Silvia, Woll, "Transhumanismus und Posthumanismus – ein Überblick, oder: Der schmale Grat zwischen Utopie und Dystopie," *Journal of New Frontiers in Spatial Concepts* 5 (2013): 43–48.

3 Stefan Herbrechter, *Posthumanismus: Eine kritische Einführung* (Darmstadt: WBG, 2009).

4 Hava Tirosh-Samuelson, "Religion," in *Post- and Transhumanism: An Introduction*, ed. Robert Ranisch and Stefan Lorenz Sorgner (Frankfurt am Main: Peter Lang, 2014), 49–71.

5 Thomas D. Philbeck, "Ontology," in *Post- and transhumanism: an introduction*, ed. Robert Ranisch and Stefan Lorenz Sorgner (Frankfurt am Main: Peter Lang, 2014), 173–183.

6 Regarding the notion of the posthuman in general, see Nicholas Gane, "Posthuman," *Theory, Culture and Society: Explorations in Critical Social Science* 23, no. 2–3 (2006): 431–434; Stefan Lorenz Sorgner, *Transhumanismus: "Die Gefährlichste Idee der Welt"!?* (Freiburg im Breisgau: Herder, 2016); Daryl J. Wennemann, "The Concept of the Posthuman: Chain of Being or Conceptual Saltus?" *Journal of Evolution and Technology* 26, no. 2 (2016): 16–30.

7 FM-2030, *Are you a Transhuman? Monitoring and Stimulating Your Personal Rate of Growth in a Rapidly Changing World* (New York: Warner Books, 1989).

8 On differentiating between the transhuman, the posthuman, and transhumanists, see Stefan Lorenz Sorgner, *Transhumanismus: "Die Gefährlichste Idee der Welt"!?* (Freiburg im Breisgau: Herder, 2016), 17; Hava Tirosh-Samuelson and Kenneth L. Mossman, "New Perspectives on Transhumanism," in *Building Better Humans? Refocusing the Debate on Transhumanism*, ed. Hava Tirosh-Samuelson and Kenneth L. Mossman (Frankfurt am Main: Peter Lang, 2012), 33.

9 Nick Bostrom, "A History of Transhumanist Thought," *Journal of Evolution and Technology* 14, no. 1 (2005): 1–25.

10 Max More, "The Philosophy of Transhumanism," in *The Transhumanist Reader: Classical and Contemporary Essays on the Science, Technology, and Philosophy of the Human Future*, ed. Max More and Natasha Vita-More (Malden, MA and Oxford: Wiley-Blackwell, 2013), 3–17.

11 Simon Young, *Designer Evolution: A Transhumanist Manifesto* (New York: Prometheus Books, 2006), 20.

12 Michael Hauskeller, "Utopia," in *Post- and Transhumanism: an Introduction*, ed. Robert Ranisch and Stefan Lorenz Sorgner (Fankfurt am Main: Peter Lang, 2014), 104.

13 Christopher Coenen and Reinhard Heil, "Historische Aspekte aktueller Menschenverbesserungsvisionen," in *Jahrbuch für Pädagogik 2014: Menschenverbesserung: Transhumanismus*, ed. Sven Kluge, Ingrid Lohmann, and Gerd Steffens (Freankfurt a. Main: Peter Lang, 2014), 45, my translation; cf. Knorr Cetina Karin, "Beyond Enlightenment: The Rise of the Culture

of Life," in The European Commission, *Modern Biology and Visions of Humanity* (Brussels: Multi-Science Publishing, 2004), 29–41; Karin Knorr Cetina, "Jenseits der Aufklärung: Die Entstehung der Kultur des Lebens," in *Bios und Zoë: Die menschliche Natur im Zeitalter ihrer technischen Reproduzierbarkeit*, ed. Martin G. Weiss (Frankfurt am Main: Suhrkamp, 2009), 55–71; James Hughes, "Contradictions from the Enlightenment Roots of Transhumanism," *Journal of Medicine and Philosophy* 35, no. 6 (2010): 622–640.

14 On the notion of control in transhumanist thinking, see also James Hughes, *Citizen Cyborg: Why Democratic Societies Must Respond to the Redesigned Human of the Future* (Boulder, CO: Westview Press, 2004), xv; Hava Tirosh-Samuelson, "Engaging Transhumanism," in *H +/–: Transhumanism and its Critics*, ed. Gregory R. Hansell and William Grassie (Philadelphia: Metanexus Institute, 2011), 46; Hava Tirosh-Samuelson and Kenneth L. Mossman, "New Perspectives on Transhumanism," in *Building Better Humans? Refocusing the Debate on Transhumanism*, ed. Hava Tirosh-Samuelson and Kenneth L. Mossman (Frankfurt am Main: Peter Lang, 2012), 29.

15 Nick Bostrom, "A History of Transhumanist Thought," *Journal of Evolution and Technology* 14, no. 1 (2005): 1–25.

16 Simon Young, *Designer Evolution: A Transhumanist Manifesto* (New York: Prometheus Books, 2006), 19.

17 Bostrom Nick, "Why I Want to be a Posthuman When I Grow Up," in *The Transhumanist Reader: Classical and Contemporary Essays on the Science, Technology, and Philosophy of the Human Future*, ed. Max More and NatashaVita-More (Malden, MA and Oxford: Wiley-Blackwell, 2013), 35; see also Max More, "The Philosophy of Transhumanism," in *The Transhumanist Reader: Classical and Contemporary Essays on the Science, Technology, and Philosophy of the Human Future*, ed. Max More and Natasha Vita-More (Malden, MA and Oxford: Wiley-Blackwell, 2013), 5; Simon Young, *Designer Evolution: A Transhumanist Manifesto* (New York: Prometheus Books, 2006), 18.

18 Martine, Rothblatt, "Mind is Deeper Than Matter: Transgenderism, Transhumanism, and the Freedom of Form," in *The Transhumanist Reader: Classical and Contemporary Essays on the Science, Technology, and Philosophy of the Human Future*, ed. Max More and Natasha Vita-More (Malden, MA and Oxford: Wiley-Blackwell, 2013), 315–326.

19 Anders Sandberg, "Morphological Freedom – Why We not Just Want It, But Need It," in *The Transhumanist Reader: Classical and Contemporary Essays on the Science, Technology, and Philosophy of the Human Future*, ed. Max More and Natasha Vita-More (Malden, MA and Oxford: Wiley-Blackwell, 2013), 56–64.

20 FM-2030, *UpWingers: A Futurist Manifesto* (New York: John Day Co, 1973); James Hughes, *Citizen Cyborg: Why Democratic Societies Must Respond to the Redesigned Human of the Future* (Boulder, CO: Westview Press, 2004), xv; James Hughes, "Report on the 2007 Interests and Beliefs Survey of the Members of the World Transhumanist Association," *World Transhumanist Association*, 2008, http://www.transhumanism.org/resources/WTASurvey2007.pdf, accessed 22 April 2017; James Hughes, "The Politics of Transhumanism and the Techno-millennial Imagination, 1626–2030," *Zygon* 47, no. 4 (2012): 757–776; James Hughes, "Politics," in *Post- and Transhumanism: An Introduction*, ed. Robert Ranisch and Stefan Lorenz Sorgner (Fankfurt am Main: Peter Lang, 2014), 133–148.

21 Ingrid Lohmann, "Menschenverbesserung mit ungewissem Ausgang," in *Jahrbuch für Pädagogik 2014: Menschenverbesserung: Transhumanismus*, ed. Sven Kluge, Ingrid Lohmann, and Gerd Steffens (Frankfurt a. Main: Peter Lang, 2014), 25.

22 Simon Young, *Designer Evolution: A Transhumanist Manifesto* (New York: Prometheus Books, 2006), 20; Sarah Chan, "Enhancement and Evolution," in *Evolution and the Future: Anthropology, Ethics, Religion*, ed. Stefan Lorenz Sorgner and Branka-Rista Jovanovic (Frankfurt am Main: Peter Lang, 2010), 49–65.

23 Christopher Coenen, "Der frühe Transhumanismus zwischen Wissenschaft und Religion," in Stefan Lorenz Sorgner (ed.), *Aufklärung und Kritik: Zeitschrift für freies Denken und humanistische Philosophie: Schwerpunkt Transhumanismus* 22, no. 3 (2015): 50–51.

24 Hans Blumenberg, *Beschreibung des Menschen* (Frankfurt am Main: Suhrkamp, 2014), 511, my translation.

25 Simon Young, *Designer Evolution: A Transhumanist Manifesto* (New York: Prometheus Books, 2006), 22.

26 Max More, "A Letter to Mother Nature," in *The Transhumanist Reader: Classical and Contemporary Essays on the Science, Technology, and Philosophy of the Human Future*, ed. Max More and NatashaVita-More (Malden, MA and Oxford: Wiley-Blackwell, 2013), 449–450; Simon Young, *Designer Evolution: A Transhumanist Manifesto* (New York: Prometheus Books, 2006), 27–29.

27 Nick Bostrom, "A History of Transhumanist Thought," *Journal of Evolution and Technology* 14, no. 1 (2005): 1–25; Max More, "The Philosophy of Transhumanism," in *The Transhumanist Reader: Classical and Contemporary Essays on the Science, Technology, and Philosophy of the Human Future*, ed. Max More and Natasha Vita-More (Malden, MA and Oxford: Wiley-Blackwell, 2013), 4; Simon Young, *Designer Evolution: A Transhumanist Manifesto* (New York: Prometheus Books, 2006), 20.

28 Dieter Birnbacher, "Neuro-enhancement," *Information Philosophie* no. 1 (2016): 18–33; Theo Boer, and Richard Fischer (eds.), *Human Enhancement: Scientific, Ethical and Theological Aspects from a European Perspective*, 2012, https://www.ceceurope.org/wp-content/uploads/2015/12/CEC-Bookonline.pdf, accessed 22 April 2017; Nick Bostrom, "In Defense of Posthuman Dignity," in *H +/- : Transhumanism and its Critics*, ed. Gregory R. Hansell and William Grassie (Philadelphia: Metanexus Institute, 2011), 55–66; Nick Bostrom, "Human Genetic Enhancements: A Transhumanist Perspective," *Journal of Value Inquiry* 37, no. 4 (2003): 493–506; Sarah Chan, "Enhancement and Evolution," in *Evolution and the Future: Anthropology, Ethics, Religion*, ed. Stefan Lorenz Sorgner and Branka-Rista Jovanovic (Frankfurt am Main: Peter Lang, 2010), 49–65; Sacha Dickel, "Der neue Mensch – ein (technik)utopisches Upgrade. Der Traum vom Human Enhancement," *Aus Politik und Zeitgeschichte (APuZ), Der Neue Mensch* 66, no. 37–38 (2016): 16–21; John Harris, "Enhancements are a Moral Obligation," in *Human Enhancement*, ed. Julian Savulescu and Nick Bostrom (Oxford: Oxford University Press, 2009), 59–70; Jan-Christoph Heilinger, *Anthropologie und Ethik des Enhancements* (Berlin and New York: De Gruyter, 2010); Nikolaus Knoepffler and Julian Savulescu (eds.), *Der neue Mensch? Enhancement und Genetik* (Freiburg: Verlag Karl Alber, 2009); Erik Parens (ed.), *Enhancing Human Traits: Ethical and Social Implications* (Washington, DC: Georgetown University Press, 2007); Ingmar Persson and Julian Savulescu, "Moral Transhumanism," *Journal of Medicine and Philosophy* 35 (2010): 656–669; Julian Savulescu and Nick Bostrom (eds.), *Human Enhancement* (Oxford: Oxford University Press, 2009); Bettina Schöne-Seifert and Davinia Talbot (eds.), *Enhancement: Die ethische Debatte* (Paderborn: Mentis, 2009); Bettina Schöne-Seifert et al. (eds.), *Neuro-enhancement: Ethik vor neuen Herausforderungen* (Paderborn: Mentis, 2009); Stefan Lorenz Sorgner, *Transhumanismus: "Die Gefährlichste Idee der Welt"!?* (Freiburg im Breisgau: Herder, 2016).

29 Cf. Rosi Braidotti, "Jenseits des Menschen: Posthumanismus," *Aus Politik und Zeitgeschichte (APuZ): Der Neue Mensch* 66, no. 37–38 (2016): 37; Rosi Braidotti, *Posthumanismus: Leben jenseits des Menschen* (Frankfurt am Main: Campus Verlag, 2014), 63; Nicole C. Karafyllis, "Das Wesen der Biofakte," in *Biofakte: Versuch über den Menschen zwischen Artefakt und Lebewesen*, ed. Nicole C. Karafyllis (Paderborn: Mentis, 2003), 23–24.

30 Thomas Damberger, "Erziehung, Bildung und pharmakologisches Enhancement," in Stefan Lorenz Sorgner (ed.), *Aufklärung und Kritik: Zeitschrift für freies Denken und humanistische Philosophie: Schwerpunkt Transhumanismus* 22, no. 3 (2015): 129, my translation.

31 Thomas Damberger, "Erziehung, Bildung und pharmakologisches Enhancement," in Stefan Lorenz Sorgner (ed.), *Aufklärung und Kritik: Zeitschrift für freies Denken und humanistische Philosophie: Schwerpunkt Transhumanismus* 22, no. 3 (2015): 129.

32 For an alternative interpretation see Erik Davis, *TechGnosis: Myth, Magic, and Mysticism in the Age of Information* (Berkeley, CA: North Atlantic Books, 1998); Boris Groys and Michael Hagemeister (eds.), *Die Neue Menschheit. Biopolitische Utopien in Russland zu Beginn des 20. Jahrhunderts* (Frankfurt am Main: Suhrkam, 2005); Boris Groys and Anton Vidokle (eds.), *Kosmismus* (Berlin: Matthes & Seitz, 2018).

33 Gerhard Engel, "Transhumanismus als Humanismus: Versuch einer Ortsbestimmung," in Stefan Lorenz Sorgner (ed.), *Aufklärung und Kritik: Zeitschrift für freies Denken und humanistische Philosophie: Schwerpunkt Transhumanismus* 22, no. 3 (2015): 28–48; Reinhard Heil, "Trans- und Posthumanismus: Eine Begriffsbestimmung," in *Endlichkeit, Medizin und Unsterblichkeit: Geschichte – Theorie – Ethik*, ed. Annette Hilt, Isabella Jordan, and Andreas Frewer (Stuttgart: Franz Steiner Verlag, 2010), 127–149.

34 Giovanni Pico della Mirandola, *De hominis dignitate*, trans. into German as *Über die Würde des Menschen* by Norbert Baumgarten, ed. and introd. August Buck, Hamburg: Felix Meiner (1990); English excerpt, "Oration on the dignity of man," http://bactra.org/Mirandola/, accessed 21 April 2017.

35 Giovanni Pico della Mirandola, *De hominis dignitate*, trans. into German as *Über die Würde des Menschen* by Norbert Baumgarten, ed. and introd. August Buck, Hamburg: Felix Meiner (1990); English excerpt, "Oration on the dignity of man," http://bactra.org/Mirandola/, accessed 21 April 2017.

36 Giovanni Pico della Mirandola, *De hominis dignitate*, trans. into German as *Über die Würde des Menschen* by Norbert Baumgarten, ed. and introd. August Buck, Hamburg: Felix Meiner (1990); English excerpt, "Oration on the dignity of man," http://bactra.org/Mirandola/, accessed 21 April 2017.

37 Giovanni Pico della Mirandola, *De hominis dignitate*, trans. into German as Über die Würde des Menschen by Norbert Baumgarten, ed. and introd. August Buck, Hamburg: Felix Meiner (1990); English excerpt, "Oration on the dignity of man," http://bactra.org/Mirandola/, accessed 21 April 2017.

38 Nick Bostrom, "Human Genetic Enhancements: A Transhumanist Perspective," *Journal of Value Inquiry* 37, no. 4 (2003): 494.

39 Bostrom Nick, "Why I Want to be a Posthuman When I Grow Up," in *The Transhumanist Reader: Classical and Contemporary Essays on the Science, Technology, and Philosophy of the Human Future*, ed. Max More and NatashaVita-More (Malden, MA and Oxford: Wiley-Blackwell, 2013), 31–32.

40 Bostrom Nick, "Why I Want to be a Posthuman When I Grow Up," in *The Transhumanist Reader: Classical and Contemporary Essays on the Science, Technology, and Philosophy of the Human Future*, ed. Max More and NatashaVita-More (Malden, MA and Oxford: Wiley-Blackwell, 2013), 28–29.

41 Nick Bostrom, "In Defense of Posthuman Dignity," in *H +/– : Transhumanism and its Critics*, ed. Gregory R. Hansell and William Grassie (Philadelphia: Metanexus Institute, 2011), 56; Michael Hauskeller, "Utopia," in *Post- and Transhumanism: an Introduction*, ed. Robert Ranisch and Stefan Lorenz Sorgner (Fankfurt am Main: Peter Lang, 2014), 107; Thomas D. Philbeck, "Ontology," in *Post- and transhumanism: an introduction*, ed. Robert Ranisch and Stefan Lorenz Sorgner (Frankfurt am Main: Peter Lang, 2014), 175; Mark Walker, "Ship of Fools: Why Transhumanism is the Best Bet to Prevent the Extinction of Civilization," in *H +/–: Transhumanism and its Critics*, ed. Gregory R. Hansell and William Grassie (Philadelphia: Metanexus Institute, 2011), 94.

42 Nicholas Gane, "Posthuman," *Theory, Culture and Society: Explorations in Critical Social Science* 23, no. 2–3 (2006): 431–434; Daryl J. Wennemann, "The Concept of the Posthuman: Chain of Being or Conceptual Saltus?" *Journal of Evolution and Technology* 26, no. 2 (2016): 16–30.

43 Stefan Lorenz Sorgner, *Transhumanismus: "Die Gefährlichste Idee der Welt"!?* (Freiburg im Breisgau: Herder, 2016); Max More, "The Philosophy of Transhumanism," in *The Transhumanist Reader: Classical and Contemporary Essays on the Science, Technology, and Philosophy of the Human Future*, ed. Max More and Natasha Vita-More (Malden, MA and Oxford: Wiley-Blackwell, 2013), 3–17; Nick Bostrom, *Superintelligence: Paths, Dangers, Strategies* (Oxford: Oxford University Press, 2014).

44 Ray Kurzweil, *The Singularity is Near: When Humans Transcend Biology* (New York: Viking, 2006), 25.

45 Susan Schneider, "Future Minds: Transhumanism, Cognitive Enhancement and the Nature of Persons," 2008, https://repository.upenn.edu/cgi/viewcontent.cgi?article=1037&context=neuroethics_pubs, accessed 21 April 2017.

46 For an overview of the current literature, see *Mind Uploading: Realizing the Goal of Substrate-Independent Minds*: http://www.minduploading.org/; see also the very helpful Wikipedia article on mind uploading: https://en.wikipedia.org/wiki/Mind_uploading; cf. Randal A. Koene, "Uploading to Substrate-independent Minds," in *The Transhumanist Reader: Classical and Contemporary Essays on the Science, Technology, and Philosophy of the Human Future*, ed. Max More and NatashaVita-More (Malden, MA and Oxford: Wiley-Blackwell, 2013), 146–156; Klaus Mathwig, "Mind Uploading – Neue Substrate für den menschlichen Geist?" n. d., http://www.jdzb.de/fileadmin/Redaktion/PDF/veroeffentlichungen/tagungsbaende/D57/13-p1184%20mathwig.pdf, accessed 22 April 2017; Ralph C. Merkle, "Uploading," in *The Transhumanist Reader: Classical and Contemporary Essays on the Science, Technology, and Philosophy of the Human Future*, ed. Max More and NatashaVita-More (Malden, MA and Oxford: Wiley-Blackwell, 2013), 157–164; Martine Rothblatt, "From Mind Loading to Mind Cloning: Gene to Meme to Beme: A Perspective on the Nature of Humanity," in *H +/–: Transhumanism and its Critics*, ed. Gregory R. Hansell and William Grassie (Philadelphia: Metanexus Institute, 2011), 112–119; Keith Wiley, *A Taxonomy and Metaphysics of Mind-uploading* (Seattle: Humanity+ Press and Alautun Press, 2014).

47 Hans Moravec, *Mind Children: The Future of Robot and Human Intelligence* (Cambridge, MA: Harvard University Press, 1988), 1.

48 Hans Moravec, "Pigs in Cyberspace," in *The Transhumanist Reader: Classical and Contemporary Essays on the Science, Technology, and Philosophy of the Human Future*, ed. Max More and NatashaVita-More (Malden, MA and Oxford: Wiley-Blackwell, 2013), 181.

49 Hans Moravec, *Mind Children: The Future of Robot and Human Intelligence* (Cambridge, MA: Harvard University Press, 1988), 116.

50 Ray Kurzweil, *The Singularity is Near: When Humans Transcend Biology* (New York: Viking, 2006), 32.

51 Ray Kurzweil, *The Singularity is Near: When Humans Transcend Biology* (New York: Viking, 2006), 38.

52 Martine Rothblatt, "From Mind Loading to Mind Cloning: Gene to Meme to Beme: A Perspective on the Nature of Humanity," in *H +/–: Transhumanism and its Critics*, ed. Gregory R. Hansell and William Grassie (Philadelphia: Metanexus Institute, 2011), 115–117.

53 Hans Moravec, *Mind Children: The Future of Robot and Human Intelligence* (Cambridge, MA: Harvard University Press, 1988), 114.

54 Hans Moravec, *Mind Children: The Future of Robot and Human Intelligence* (Cambridge, MA: Harvard University Press, 1988), 115–116. Cf. Marvin Minsky, *The Society of Mind* (New York: Touchstone Books, 1986).

55 James Hughes, "The Future of Death: Cryonics and the Telos of Liberal Individualism," *Journal of Evolution and Technology* 6, no. 1 (2001): 1–23; Stefan Lorenz Sorgner, *Transhumanismus: "Die Gefährlichste Idee der Welt"!?* (Freiburg im Breisgau: Herder, 2016), 67; and for a critical discussion, see Oliver Oliber Krüger, "Die Vervollkommnung des Menschen: Tod und Unsterblichkeit im Posthumanismus und Transhumanismus," *Eurozine*, August 17, 2007, https://www.eurozine.com/die-vervollkommnung-des-menschen/, accessed 22 April 2017.

56   Hannah Arendt, "Natur und Geschichte," in idem, *Zwischen Vergangenheit und Zukunft: Übungen im politischen Denken I*, ed. Ursula Ludz, 3rd edn. (Munich: Piper, 2015), 67.

57   Hannah Arendt, *Vita activa, oder vom tätigen Leben*, 8th edn. (Munich: Piper, 2010), originally published in English as *The Human Condition* (Chicago: University of Chicago Press, 1958), 219.

REFERENCES

Arendt, Hannah, *Vita activa, oder vom tätigen Leben*, 8th edn. (Munich: Piper, 2010), originally published in English as *The Human Condition* (Chicago: University of Chicago Press, 1958).

Arendt, Hannah, "Natur und Geschichte," in idem, *Zwischen Vergangenheit und Zukunft: Übungen im politischen Denken I*, ed. Ursula Ludz, 3rd edn. (Munich: Piper, 2015), 54–79.

Birnbacher, Dieter, "Posthumanity, transhumanism, and human nature," in *Medical Enhancement and Posthumanity*, ed. Bert Gordijn and Ruth Chadwick (Berlin and New York: Springer, 2008), 95–106.

Birnbacher, Dieter, "Transhumanismus – Provokation oder Trivialität?" in Stefan Lorenz Sorgner (ed.), *Aufklärung und Kritik: Zeitschrift für freies Denken und humanistische Philosophie: Schwerpunkt Transhumanismus* 22, no. 3 (2015): 130–151.

Birnbacher, Dieter, "Neuro-enhancement," *Information Philosophie* no. 1 (2016): 18–33.

Blumenberg, Hans, *Beschreibung des Menschen* (Frankfurt am Main: Suhrkamp, 2014).

Boer, Theo, and Richard Fischer (eds.), *Human Enhancement: Scientific, Ethical and Theological Aspects from a European Perspective*, 2012, https://www.ceceurope.org/wp-content/uploads/2015/12/CEC-Bookonline.pdf.

Bostrom, Nick, "Human Genetic Enhancements: A Transhumanist Perspective," *Journal of Value Inquiry* 37, no. 4 (2003): 493–506.

Bostrom, Nick, "A History of Transhumanist Thought," *Journal of Evolution and Technology* 14, no. 1 (2005): 1–25.

Bostrom, Nick, "In Defense of Posthuman Dignity," in *H +/– : Transhumanism and its Critics*, ed. Gregory R. Hansell and William Grassie (Philadelphia, PA: Metanexus Institute, 2011), 55–66.

Bostrom, Nick, "Why I Want to be a Posthuman When I Grow Up," in *The Transhumanist Reader: Classical and Contemporary Essays on the Science, Technology, and Philosophy of the Human Future*, ed. Max More and Natasha Vita-More (Malden, MA and Oxford: Wiley-Blackwell, 2013), 28–53.

Bostrom, Nick, *Superintelligence: Paths, Dangers, Strategies* (Oxford: Oxford University Press, 2014).

Braidotti, Rosi, *Posthumanismus: Leben jenseits des Menschen* (Frankfurt am Main: Campus Verlag, 2014).

Braidotti, Rosi, "Jenseits des Menschen: Posthumanismus," *Aus Politik und Zeitgeschichte (APuZ): Der Neue Mensch* 66, no. 37–38 (2016): 33–38.

Chan, Sarah, "Enhancement and Evolution," in *Evolution and the Future: Anthropology, Ethics, Religion*, ed. Stefan Lorenz Sorgner and Branka-Rista Jovanovic (Frankfurt am Main: Peter Lang, 2010), 49–65.

Coenen, Christopher, "Der frühe Transhumanismus zwischen Wissenschaft und Religion," in Stefan Lorenz Sorgner (ed.), *Aufklärung und Kritik: Zeitschrift für freies Denken und humanistische Philosophie: Schwerpunkt Transhumanismus* 22, no. 3 (2015): 49–61.

Coenen, Chrisstopher, and Reinhard Heil, "Historische Aspekte aktueller Menschenverbesserungsvisionen," in *Jahrbuch für Pädagogik 2014: Menschenverbesserung: Transhumanismus*, ed. Sven Kluge, Ingrid Lohmann, and Gerd Steffens (Freankfurt a. Main: Peter Lang, 2014), 35–49.

Damberger, Thomas, "Erziehung, Bildung und pharmakologisches Enhancement," in Stefan Lorenz Sorgner (ed.), *Aufklärung und Kritik: Zeitschrift für freies Denken und humanistische Philosophie: Schwerpunkt Transhumanismus* 22, no. 3 (2015): 129–139.

Davis, Erik, *TechGnosis: Myth, Magic, and Mysticism in the Age of Information* (Berkeley, CA: North Atlantic Books, 1998).

Dickel, Sascha, "Der neue Mensch – ein (technik)utopisches Upgrade. Der Traum vom Human Enhancement," *Aus Politik und Zeitgeschichte (APuZ), Der Neue Mensch* 66, no. 37–38 (2016): 16–21.

Engel, Gerhard, "Transhumanismus als Humanismus: Versuch einer Ortsbestimmung," in Stefan Lorenz Sorgner (ed.), *Aufklärung und Kritik: Zeitschrift für freies Denken und humanistische Philosophie: Schwerpunkt Transhumanismus* 22, no. 3 (2015): 28–48.

FM-2030, *UpWingers: A Futurist Manifesto* (New York: John Day Co, 1973).

FM-2030, *Are You a Transhuman? Monitoring and Stimulating Your Personal Rate of Growth in a Rapidly Changing World* (New York: Warner Books, 1989).

Gane, Nicholas, "Posthuman," *Theory, Culture and Society: Explorations in Critical Social Science* 23, no. 2–3 (2006): 431–434.

Garreau, Joel, *Radical Evolution: The Promise and Peril of Enhancing Our Minds, our Bodies – And What it Means to be Human* (New York: Broadway Books, 2005).

Groys, Boris, and Michael Hagemeister (eds.), *Die Neue Menschheit. Biopolitische Utopien in Russland zu Beginn des 20. Jahrhunderts* (Frankfurt am Main: Suhrkam, 2005).

Groys, Boris, and Anton Vidokle (eds.), *Kosmismus* (Berlin: Matthes & Seitz, 2018).

Hansell, Greogry R., and William Grassie (eds.), *H +/–: Transhumanism and its Critics* (Philadelphia: Metanexus Institute, 2011).

Harris, John, "Enhancements are a Moral Obligation," in *Human Enhancement,* ed. Julian Savulescu and Nick Bostrom (Oxford: Oxford University Press, 2009), 59–70.

Hauskeller, Michael, "Utopia," in *Post- and Transhumanism: An Introduction,* ed. Robert Ranisch and Stefan Lorenz Sorgner (Fankfurt am Main: Peter Lang, 2014), 101–108.

Heil, Reinhard, "Trans- und Posthumanismus: Eine Begriffsbestimmung," In *Endlichkeit, Medizin und Unsterblichkeit: Geschichte – Theorie – Ethik,* ed. Annette Hilt, Isabella Jordan and Andreas Frewer (Stuttgart: Franz Steiner Verlag, 2010), 127–149.

Heilinger, Jan-Christoph, *Anthropologie und Ethik des Enhancements* (Berlin and New York: De Gruyter, 2010).

Herbrechter, Stefan, *Posthumanismus: Eine kritische Einführung* (Darmstadt: WBG, 2009).

Hughes, James, "The Future of Death: Cryonics and the Telos of Liberal Individualism," *Journal of Evolution and Technology* 6, no. 1 (2001): 1–23.

Hughes, James, *Citizen Cyborg: Why Democratic Societies Must Respond to the Redesigned Human of the Future* (Boulder, CO: Westview Press, 2004).

Hughes, James, "Report on the 2007 Interests and Beliefs Survey of the Members of the World Transhumanist Association," *World Transhumanist Association,* 2008, https://www.transhumanism.org/resources/WTASurvey2007.pdf.

Hughes, James, "Contradictions from the Enlightenment Roots of Transhumanism," *Journal of Medicine and Philosophy* 35, no. 6 (2010): 622–640.

Hughes, James, "The Politics of Transhumanism and the Techno-millennial Imagination, 1626–2030," *Zygon* 47, no. 4 (2012): 757–776.

Hughes, James, "Politics," in *Post- and transhumanism: An introduction,* ed. Robert Ranisch and Stefan Lorenz Sorgner (Fankfurt am Main: Peter Lang, 2014), 133–148.

Irrgang, Bernhard, *Posthumanes Menschsein? Künstliche Intelligenz, Cyberspace, Roboter, Cyborgs und Designer-Menschen – Anthropologie des künstlichen Menschen im 21. Jahrhundert* (Stuttgart: Franz Steiner Verlag, 2005).

Karafyllis, Nicole C., "Das Wesen der Biofakte," in *Biofakte: Versuch über den Menschen zwischen Artefakt und Lebewesen,* ed. Nicole C. Karafyllis (Paderborn: Mentis, 2003), 11–26.

Kluge, Sven, Ingrid Lohmann, and Gerd Steffens (eds.), *Jahrbuch für Pädagogik 2014: Menschenverbesserung: Transhumanismus* (Frankfurt a. Main: Peter Lang, 2014).

Knoepffler, Nikolaus, and Julian Savulescu (eds.), *Der neue Mensch? Enhancement und Genetik* (Freiburg: Verlag Karl Alber, 2009).

Knorr Cetina, Karin, "Beyond Enlightenment: The Rise of the Culture of Life," in The European Commission, *Modern Biology and Visions of Humanity* (Brussels: Multi-Science Publishing, 2004), 29–41.

Knorr Cetina, Karin, "Jenseits der Aufklärung: Die Entstehung der Kultur des Lebens," in *Bios und Zoë: Die menschliche Natur im Zeitalter ihrer technischen Reproduzierbarkeit*, ed. Martin G. Weiss (Frankfurt am Main: Suhrkamp, 2009), 55–71.

Koene, Randal A., "Uploading to Substrate-independent Minds," in *The Transhumanist Reader: Classical and Contemporary Essays on the Science, Technology, and Philosophy of the Human Future*, ed. Max More and Natasha Vita-More (Malden, MA and Oxford: Wiley-Blackwell, 2013), 146–156.

Krüger, Oliver, "Die Vervollkommnung des Menschen: Tod und Unsterblichkeit im Posthumanismus und Transhumanismus," *Eurozine*, August 17, 2007, https://www.eurozine.com/die-vervollkommnung-des-menschen/.

Kurzweil, Ray, *The Singularity is Near: When Humans Transcend Biology* (New York: Viking, 2006).

Loh, Janina, *Trans- und Posthumanismus zur Einführung* (Hamburg: Junius, 2018).

Lohmann, Ingrid, "Menschenverbesserung mit ungewissem Ausgang," in *Jahrbuch für Pädagogik 2014: Menschenverbesserung: Transhumanismus*, ed. Sven Kluge, Ingrid Lohmann, and Gerd Steffens (Frankfurt a. Main: Peter Lang, 2014), 17–31.

Mathwig, Klaus, "Mind uploading – Neue Substrate für den menschlichen Geist?" n. d., https://www.jdzb.de/fileadmin/Redaktion/PDF/veroeffentlichungen/tagungsbaende/D57/13-p1184%20mathwig.pdf.

Merkle, Ralph C., "Uploading," in *The Transhumanist Reader: Classical and Contemporary Essays on the Science, Technology, and Philosophy of the Human Future*, ed. Max More and Natasha Vita-More (Malden, MA and Oxford: Wiley-Blackwell, 2013), 157–164.

Minsky, Marvin, *The Society of Mind* (New York: Touchstone Books, 1986).

Moravec, Hans, *Mind Children: The Future of Robot and Human Intelligence* (Cambridge, MA: Harvard University Press, 1988).

Moravec, Hans, "Pigs in Cyberspace," in *The Transhumanist Reader: Classical and Contemporary Essays on the Science, Technology, and Philosophy of the Human Future*, ed. Max More and Natasha Vita-More (Malden, MA and Oxford: Wiley-Blackwell, 2013), 177–181.

More, Max, "The Philosophy of Transhumanism," in *The Transhumanist Reader: Classical and Contemporary Essays on the Science, Technology, and Philosophy of the Human Future*, ed. Max More and Natasha Vita-More (Malden, MA and Oxford: Wiley-Blackwell, 2013), 3–17.

More, Max, "A Letter to Mother Nature," in *The Transhumanist Reader: Classical and Contemporary Essays on the Science, Technology, and Philosophy of the Human Future*, ed. Max More and Natasha Vita-More (Malden, MA and Oxford: Wiley-Blackwell, 2013), 449–450.

More, Max, and Natasha Vita-More (eds.), *The Transhumanist Reader: Classical and Contemporary Essays on the Science, Technology, and Philosophy of the Human Future* (Malden, MA and Oxford: Wiley-Blackwell, 2013).

Niemeyer, Christian, "Vom Transhumanismus zurück zur Transformation, altdeutsch: 'Verwandlung,'" in Stefan Lorenz Sorgner (ed.), *Aufklärung und Kritik: Zeitschrift für freies Denken und humanistische Philosophie: Schwerpunkt Transhumanismus* 22, no. 3 (2015): 130–144.

Parens, Erik, (ed.), *Enhancing Human Traits: Ethical and Social Implications* (Washington, DC: Georgetown University Press, 2007).

Persson, Ingmar, and Julian Savulescu, "Moral Transhumanism," *Journal of Medicine and Philosophy* 35, (2010): 656–669.

Philbeck, Thomas D., "Ontology," in *Post- and Transhumanism: An Introduction*, ed. Robert Ranisch and Stefan Lorenz Sorgner (Frankfurt am Main: Peter Lang, 2014), 173–183.

Pico della Mirandola, Giovanni, *De hominis dignitate*, trans. into German as *Über die Würde des Menschen* by Norbert Baumgarten, ed. and introd. August Buck (Hamburg: Felix Meiner, 1990); English excerpt, "Oration on the Dignity of Man," https://bactra.org/Mirandola/, accessed 21 April 2017.

Ranisch, Robert, "Morality," in *Post- and Transhumanism: An Introduction,* ed. Robert Ranisch and Stefan Lorenz Sorgner (Frankfurt am Main: Peter Lang, 2014), 149–172.

Ranisch, Robert, and Stefan Lorenz Sorgner (eds.), *Post- and Transhumanism: An Introduction* (Frankfurt am Main: Peter Lang, 2014).

Rothblatt, Martine, "From Mind Loading to Mind Cloning: Gene to Meme to Beme: A Perspective on the Nature of Humanity," In *H +/−: Transhumanism and its Critics,* ed. Gregory R. Hansell and William Grassie (Philadelphia, PA: Metanexus Institute, 2011), 112–119.

Rothblatt, Martine, "Mind is Deeper Than Matter: Transgenderism, Transhumanism, and the Freedom of Form," in *The Transhumanist Reader: Classical and Contemporary Essays on the Science, Technology, and Philosophy of the Human Future,* ed. Max More and Natasha Vita-More (Malden, MA and Oxford: Wiley-Blackwell, 2013), 315–326.

Sandberg, Anders, "Morphological Freedom – Why We not Just Want It, But Need It," in *The Transhumanist Reader: Classical and Contemporary Essays on the Science, Technology, and Philosophy of the Human Future,* ed. Max More and Natasha Vita-More (Malden, MA and Oxford: Wiley-Blackwell, 2013), 56–64.

Savulescu, Julian, and Nick Bostrom (eds.), *Human Enhancement* (Oxford: Oxford University Press, 2009).

Schneider, Susan, "Future Minds: Transhumanism, Cognitive Enhancement and the Nature of Persons," 2008, https://repository.upenn.edu/cgi/viewcontent.cgi?article=1037&context=neur oethics_pubs.

Schöne-Seifert, Bettina, and Davinia Talbot (eds.), *Enhancement: Die ethische Debatte* (Paderborn: Mentis, 2009).

Schöne-Seifert, Bettina, Davinia Talbot, Uwe Opolka, and Johann S. Ach (eds.), *Neuro-enhancement: Ethik vor neuen Herausforderungen* (Paderborn: Mentis, 2009).

Sorgner, Stefan Lorenz (ed.), "Aufklärung und Kritik: Zeitschrift für freies Denken und humanistische Philosophie: Schwerpunkt Transhumanismus," 22, no. 3 (2015).

Sorgner, Stefan Lorenz, *Transhumanismus: "Die Gefährlichste Idee der Welt"!?* (Freiburg im Breisgau: Herder, 2016).

Tirosh-Samuelson, Hava, "Engaging Transhumanism," in *H +/−: Transhumanism and its Critics,* ed Gregory R. Hansell and William Grassie (Philadelphia, PA: Metanexus Institute, 2011), 19–53.

Tirosh-Samuelson, Hava, "Religion," in *Post- and Transhumanism: An Introduction,* ed. Robert Ranisch and Stefan Lorenz Sorgner (Frankfurt am Main: Peter Lang, 2014), 49–71.

Tirosh-Samuelson, Hava, and Kenneth L. Mossman, "New Perspectives on Transhumanism," in *Building Better Humans? Refocusing the Debate on Transhumanism,* ed. Hava Tirosh-Samuelson and Kenneth L. Mossman (Frankfurt am Main: Peter Lang, 2012), 29–52.

Walker, Mark, "Ship of Fools: Why Transhumanism is the Best Bet to Prevent the Extinction of Civilization," in *H +/−: transhumanism and its critics,* ed. Gregory R. Hansell and William Grassie (Philadelphia, PA: Metanexus Institute, 2011), 94–111.

Wennemann, Daryl J., "The Concept of the Posthuman: Chain of Being or Conceptual Saltus?" *Journal of Evolution and Technology* 26, no. 2 (2016): 16–30.

Wiley, Keith, *A Taxonomy and Metaphysics of Mind-Uploading* (Seattle: Humanity+ Press and Alautun Press, (2014).

Woll, Silvia, "Transhumanismus und Posthumanismus – ein Überblick, oder: Der schmale Grat zwischen Utopie und Dystopie," *Journal of New Frontiers in Spatial Concepts* 5 (2013): 43–48.

CHAPTER 4

Corporate Spies

Industrial Cyber Espionage and the Obligation to Prevent Trans-Boundary Harm

Russell Buchan

INTRODUCTION

There is no agreed upon definition of espionage under international law.[1] This being said, the defining feature of espionage is the theft of confidential information and this practice takes different forms depending upon the actors involved and the type of information targeted.[2] Industrial espionage involves companies stealing confidential business information belonging to other companies and it usually exhibits a trans-boundary element, that is, companies stealing trade secrets from their foreign competitors.

Industrial espionage is not a new phenomenon. However, the prevalence of this activity has increased dramatically since the dawn of cyberspace and there are three main reasons for this: first, companies use cyberspace to store colossal amounts of confidential data, making it a fertile environment for corporate spies to target; second, the virtual nature of cyberspace means that corporate spies can hide or obfuscate their identity through various types of anonymizing software or techniques, making it very unlikely that their true identity will be revealed; and third, the remote nature of industrial cyber espionage means that corporate spies can steal trade secrets safe in the knowledge that, even if their identity is uncovered by law enforcement agencies, the chances of them being extradited to a foreign jurisdiction to face charges of cyber spying are very low.

The international society has implemented various initiatives to combat industrial cyber espionage. For example, a number of states have signed bilateral agreements with China which declare that they will not engage in or support the theft of confidential business

DOI: 10.1201/9781003226406-5

information from foreign companies.³ Moreover, in 2015, G20 leaders issued a joint communiqué affirming

> [N]o country should conduct or support ICT-enabled theft of intellectual property, including trade secrets or other confidential business information, with the intent of providing competitive advantages to companies or commercial sectors.⁴

Importantly, these initiatives do not impose *binding* international legal obligations upon states; instead, they are soft law measures which set standards of acceptable conduct in cyberspace that states *should* follow. The objective of this chapter is to explore whether international law can be harnessed to regulate industrial cyber espionage, and it looks specifically at the obligation to prevent trans-boundary harm.

In pursuit of this objective, this chapter is structured as follows. Section 2 examines the impact of industrial cyber espionage on the national security of victim states and in doing so demonstrates the need for international legal regulation of this activity. Section 3 first identifies the obligation to prevent trans-boundary as a principle of customary international law; second, it demonstrates that this obligation applies in cyberspace; and third, it examines whether this obligation applies to industrial cyber espionage. Section 4 analyses the extent to which this obligation requires states to prevent industrial cyber espionage. Section 5 offers concluding remarks.

THE ECONOMIC DIMENSION OF NATIONAL SECURITY

Historically, states defined their national security in narrow terms – national security was achieved where their territorial integrity and political independence was protected from external threat. Since at least the end of the Cold War, the maintenance of national security has been defined more broadly and includes not just the preservation of the physical and political features of statehood but also the social, cultural and economic systems that states embrace.⁵ With regard to economic matters, the breakup of the Soviet Union during the 1990s revealed the particularly close nexus between a state's economic security and its national security: the Soviet Union's demise was not so much due to its military inferiority but instead its failed economic policies.

The economic prosperity of companies is critical to the national security of states.⁶ In short, successful companies retain and create jobs, pay taxes and attract national and foreign investment. It goes without saying that, for companies to be successful, they must keep certain information secret. Many examples spring to mind in this regard: customer details, employees' salaries, research and development processes, future marketing campaigns, etc. Where a foreign competitor steals this confidential information, the competitive edge of the victim company is blunted and, ultimately, its ability to generate profits and income is jeopardised. Given that national security is intimately linked to economic security, it follows that, where a company's economic success is compromised, the national security of the host state is also threatened.⁷ That industrial espionage has exploded in the cyber era only serves to raise the costs of this activity on national security. Indeed, it is for this reason that industrial cyber espionage is often presented as not just a threat to the national security of the victim state but also to the stability of the international economic order more generally.[8]

States and companies have developed various strategies to combat industrial cyber espionage. Companies, for example, deploy a wide range of passive and active cyber defenses to protect the confidentiality of their data: multiuser authentication, firewalls, encryption, etc. Yet, perpetrators of malicious cyber activity tend to devise innovative methods to circumvent cyber defenses and it is often the case that legitimate users of cyberspace are playing catch up with malicious cyber actors.[9]

States also adopt national laws to counter industrial cyber espionage and this usually takes the form of criminal sanctions. The use of national criminal law to deter and punish industrial cyber espionage is likely to be at its most effective when the spy is located within the same geographical space as the victim company. However, where the act of industrial espionage is conducted remotely through cyberspace, it is unlikely that the corporate spy can be brought within the jurisdiction of the state which plays host to the victim company. While a prosecution *in absentia* is possible, even if there is a conviction, the likelihood of the spy being extradited to the prosecuting state and punished is remote, thus diminishing the effectiveness of national criminal law.

Where industrial cyber espionage has a trans-boundary dimension (insofar as the perpetrator is located in a different jurisdiction to the victim company), this raises the question of whether international law can be called upon to regulate this practice. There are various reasons why readers may be understandably skeptical as to whether international law can play a useful role in deterring and punishing industrial cyber espionage. However, as we shall see, these concerns are overstated.

First, some may question the utility of international law given that this is a state-based system and, as we have seen, industrial cyber espionage is carried out by non-state actors against non-state actors, namely, companies against companies. Notwithstanding the state-centrist structure of international law, it remains the case that companies are located within states and thus fall under their jurisdiction. As we will observe as this chapter progresses, international law places obligations upon states to ensure that their territory is not used in a manner detrimental to the rights and interests of other states. In this way, international law is able to *indirectly* regulate industrial cyber espionage.[10]

Second, commentators may question the use of international law given that it lacks a centralized authority that is capable of enforcing its rules. However, where it can be demonstrated that a rule of international law has been violated – for example, when a state fails to prevent actors within its jurisdiction from harming the legal rights of other states – this inflicts significant reputational (and thus political) costs upon the wrongdoing state. Put differently, it is in the interests of states to comply with international law.[11] Moreover, the international legal system permits states to take countermeasures against wrongdoing states, which are unilateral self-help measures that are designed to pressurize a wrongdoing state into complying with its international legal commitments.[12]

For these reasons, international law can provide a useful regulatory mechanism and the remainder of this chapter will therefore assesses whether there are any rules of international law that can be called upon to tackle the threat posed by industrial cyber espionage.

CUSTOMARY LAW AND THE OBLIGATION TO PREVENT TRANS-BOUNDARY HARM

Article 38(1)(b) of the Statute of the International Court of Justice 1945 identifies custom as a source of international law. Customary law forms on the back of widespread and representative state practice that is conducted out of a sense of legal obligation.[13]

It is well established that customary law imposes an obligation upon states to prevent their territory from being used to cause injury to the legal rights of other states. This customary law was recognized by the ICJ in the *Corfu Channel* case, which involved UK warships hitting mines as they passed through the territorial waters of Albania (the Corfu Channel). While the Court was unable to determine that Albania had actually laid the mines itself, it did find Albania responsible under international law for its failure to take reasonable measures to ensure that vessels could conduct innocent passage through its waters free from hazards. In the enduring words of the Court, it is "well-recognized" that states are subject to a customary law obligation "not to allow knowingly its territory to be used for acts contrary to the rights of other States."[14]

The *Corfu Channel* case was decided in 1949 and the question that arises is whether states are under an obligation to prevent harmful activities emanating from their cyber infrastructure and which interfere with the legal rights of other states. The obligation to prevent trans-boundary harm that was identified in the *Corfu Channel* case is a general rule of customary law and encompasses all domains falling within a state's sovereign jurisdiction: land, territorial waters, national airspace, and cyberspace.

The UN Group of Governmental Experts (GGE) represents a group of states that identifies and examines the acceptable standards of conduct in cyberspace. In its 2015 report, the GGE determined that states "should" prevent their cyber infrastructure from being used to cause harm to other states.[15] The use of the word "should" suggests that the GGE does not regard international law as imposing a mandatory obligation upon states to prevent trans-boundary harm, or at least that, if this obligation is part of customary law, it is not transposable to the cyber setting. This is an erroneous interpretation of international law. The obligation to prevent trans-boundary harm has been an established rule of customary law for many decades, if not longer. It cannot be the case that a report in 2015 by a limited number of states can wash away its legal status, especially as there may have been many reasons why a political body such as the GGE chose the language it did. Moreover, since 2015, several states (including those who had a seat on the 2015 GGE) have come out in favor of a rule of customary law *requiring* states to prevent their cyber infrastructure from being used to cause harm to other states.[16]

The immediate question is: what type of harmful conduct must states prevent emanating from their territory? The mainstream view is that states must prevent the occurrence of conduct which, if attributed to them under the law on state responsibility, would violate a rule of international law.[17] This approach is logical because states cannot be expected as a matter of international law to prevent activities emanating from their territories that would be lawful under international law if they had committed those acts themselves. The critical issue, then, is whether acts of industrial cyber espionage would, if conducted by a state, violate a rule of international law. Two options present themselves: Article 10*bis*(1) of the Paris Convention for the Protection of Industrial Property 1967 and the principle of territorial sovereignty under customary law, each of which will be considered in turn.

Paris Convention 1967

For those states that are party to the Paris Convention, Article 10*bis*(1) requires them to "assure to nationals" of other states parties "effective protection against unfair competition." In this context, nationals include natural as well as legal persons; specifically, individuals granted nationality under the domestic law of a state party to the Paris Convention and companies incorporated under the domestic law of a state party.[18] Exceptionally, nationals of states not party to the Paris Convention are entitled to the protections afforded by this treaty when individuals are "domiciled" within the territory of a state party and when companies operate "real and effective industrial or commercial establishments" within such territory.[19]

As noted above, industrial cyber espionage is usually targeted against foreign rivals. This raises the question of whether Article 10*bis*(1) imposes an obligation upon states parties to protect nationals from unfair competition where they are located outside of their territory. Whether a treaty imposes extraterritorial obligations upon states parties depends on the construction of the provision in question and an appraisal of the object and purpose of the treaty.[20] If we look to Article 10*bis* specifically and the Paris Convention more generally, there is nothing within this provision or this agreement to indicate that states parties must only protect nationals from unfair competition when they are located within their own territory. In fact, the language employed by Article 10*bis* is broad and suggests that states parties must afford nationals protection against unfair competition regardless of where they are located: Article 10*bis*(1) requires states parties to assure to nationals effective protection against unfair competition in a general sense and Article 10*bis*(2) explains that "[a]ny act of competition contrary to honest practices" amounts to unfair competition.

This interpretation of Article 10*bis*(1) does not mean that states parties must protect nationals located within foreign territories from all acts of unfair competition. A useful analogy can be drawn with international human rights law. The jurisprudence of international human rights bodies reveals that states owe human rights obligations extraterritorially where they exercise their authority and control over individuals.[21] In this way, the extraterritorial obligation is cast in negative terms: when states act, they must do so compliantly with recognized human rights standards. A similar argument can be made with regard to the Paris Convention: when states parties exercise their authority and control over nationals, they are under a negative obligation to refrain from engaging in conduct that amounts to an act of unfair competition within the meaning of Article 10*bis*(1).

As said, Article 10*bis* prohibits acts of "unfair competition" against nationals. Article 10*bis*(2) defines unfair competition as "[a]ny act of competition contrary to honest practices in industrial or commercial matters." Examples of acts of unfair competition are listed in Article 10*bis*(3) and, while this provision does not explicitly state that the theft of trade secrets constitutes an act of unfair competition, its identification of examples that are "in particular … prohibited" suggests that list is intended to be indicative rather than exhaustive.

The notion of unfair competition comprises two distinct elements: first, an act of *competition* and, second, that it is *unfair*. Whether conduct amounts to an act of competition depends on whether the impugned act impairs the competitive opportunities of the targeted national.[22] The answer to this question is affirmative in the context of industrial cyber espionage given that, when an actor steals another company's trade secrets, the victim company's competitive opportunities are clearly undermined. It also clear that industrial cyber espionage qualifies as an act of *unfair* competition on the basis that the theft of another company's confidential business information is an intrinsically dishonest commercial practice.

The Principle of Territorial Sovereignty

Industrial cyber espionage – if committed by a state – would also violate international law insofar as it would run into conflict with the principle of territorial sovereignty. In recent years, an interesting debate has emerged among international lawyers as to whether the principle of territorial sovereignty is a recognized rule of customary law. This debate was instigated in May 2018 when the UK determined that sovereignty is a political principle of international relations rather than a binding rule of international law.[23] The UK's view has been criticized given that state practice and the jurisprudence of international tribunals prior to 2018 strongly indicates that territorial sovereignty is a rule of international law.[24] Moreover, and perhaps more importantly, since 2018 a number of other states have come out against the UK position and in doing so have endorsed the "sovereignty as a rule" position.[25]

The principle of territorial sovereignty protects the right of states to perform governmental functions within their territory.[26] Accordingly, states are prohibited from interfering with the right of other states to conduct governmental functions. What constitutes a governmental function depends on the political constitution of each state – in short, different states allocate different functions to their governments. However, certain core governmental functions can be identified. For example, a common function across governments is the right to determine who enters and who leaves their territory.

It is widely recognized that states exercise territorial sovereignty over the cyber infrastructure located within their territory and that their sovereignty also extends to the computer networks and systems supported by that infrastructure.[27] This proposition stands regardless of whether the cyber infrastructure, networks or systems are publicly or privately operated – the decisive issue is that the cyber infrastructure is located within the jurisdiction (and thus falls under the sovereignty) of the host state.[28] The majority of the Tallinn Manual 2.0 experts concluded that remote cyber operations violate the rule of territorial sovereignty only where they produce harmful effects within the target state.[29] In particular, these experts averred that remote cyber operations depriving systems and networks of their functionality, or where they damage or delete data, are examples of cyber conduct that produce sufficiently harmful effects upon a victim state to occasion a violation of the rule of territorial sovereignty. Clearly, this approach does not regard acts of cyber espionage as violating the territorial sovereignty of a victim state.[30]

In the physical world, it is well established that, where a state trespasses into the territory of another state, such conduct is unlawful irrespective of whether it produces damage – to be clear, it is the non-consensual intrusion into the territory of another state that breaches the rule of territorial sovereignty. In the *Nicaragua* case, for example, the ICJ was clear that the US's reconnaissance flights into Nicaraguan airspace breached the sovereignty of the latter state.[31]

Determining what types of cyber operations – and in particular, whether cyber *espionage* operations – violate the territorial sovereignty of a state hinges on state practice. The question here is what have states said and done in this context? When do states regard their sovereignty as being breached by a remote cyber operation? As a general matter, states have been reluctant to express publicly their views as to how international law applies to cyberspace, except for perhaps broad and abstract statements confirming the application of core rules to this domain.[32] This being said, in an important development in 2019, France adopted the view that any non-consensual cyber operation against its cyber infrastructure violates its territorial sovereignty:

> Any unauthorised penetration by a State of French systems or any production of effects on French territory via a digital vector may constitute, at the least, a breach of sovereignty.[33]

Iran[34] and Switzerland[35] have also adopted this interpretation of the principle of territorial sovereignty.

With regard to cyber espionage specifically, states are especially unwilling to publicly articulate their view because of the "policy of silence" that typically accompanies the work of intelligence agencies.[36] Nonetheless, the revelations by Edward Snowden in 2013 that the US had been engaged in a massive global cyber surveillance campaign provoked several states into declaring that the US's actions were internationally wrongful and were so on the basis that they violated the principle of territorial sovereignty. For example, MERCOSUR – a South American trading bloc representing Argentina, Bolivia, Brazil, Uruguay and Venezuela – submitted a *note verbale* to the UN Secretary-General "[c]ondemning the acts of espionage carried out by intelligence agencies of the United States of America... [which] constitute unacceptable behaviour that violates our sovereignty."[37] Separately, the Foreign Minister of Venezuela explained before the Security Council that "we reject the actions of global espionage carried out by the Government of the United States, which undermine the sovereignty of States" and also called on the UN to "punish and condemn this violation of international law."[38]

China's reaction to the Snowden disclosures echoes the views of South American states. In a speech delivered at the National Congress of Brazil, President Xi Jinping expressed that "[n]o matter how developed a country's Internet technology is, it just cannot violate the information sovereignty of other countries."[39]

According to this interpretation of the rule of territorial sovereignty, acts of industrial cyber espionage would – if attributed to a state – breach the territorial sovereignty of the state that plays host to the victim company.[40] The immediate question is whether the obligation to prevent trans-boundary harm applies to all conduct which, if attributable to a state, would violate international law, or whether the obligation only applies to conduct giving rise to sufficiently serious adverse consequences for the victim state.

International jurisprudence seems fairly clear that a *de minimis* threshold is built into the obligation to prevent trans-boundary, that is, the obligation is activated only where the conduct emanating from its territory produces sufficiently serious adverse consequences for the victim state.[41] In the context of industrial cyber espionage, this means that sporadic acts of espionage that result in the theft of fairly innocuous confidential business information would not trigger the state's obligation to prevent. But a widespread and persistent campaign of industrial cyber espionage on behalf of certain companies may be sufficiently serious to demand action from the state. Similarly, isolated acts of industrial cyber espionage targeting highly valuable confidential business information may cross the *de minimis* threshold.

Moving forward, it is important to point out that the obligation to prevent trans-boundary harm comprises two essential components: first, an absolute duty to establish governmental infrastructure; and second, a relative duty to use this infrastructure reasonably to prevent acts giving rise to the threat of trans-boundary harm.[42]

THE OBLIGATION TO PREVENT TRANS-BOUNDARY HARM

Prevention as an Obligation of Result

The legal criteria for statehood under international law are well established: to become states, political communities must possess a government that exercises effective control over a population of people within a defined territory.[43] Inherent to the notion of effective government is the requirement that states are able to organize their internal affairs in a manner such that they can meet their duties under international law. This means that states must acquire the capacity to prevent conduct emanating from their territory that is contrary to the international legal rights of other states, which includes the specific duties to detect, counteract, and mitigate harmful activity, to investigate and punish those responsible for engaging in this conduct, and to provide restitution for victims where injury is caused.[44]

The obligation upon states to acquire the capacity to prevent trans-boundary harm requires them to possess a legal and administrative system that is sufficient to enable them to prevent acts emanating from their territory that interfere with the legal rights of other states.[45] The duty upon states to establish laws and institutions capable of preventing trans-boundary harm is therefore an obligation of result and this means that states are under an absolute duty to implement them.[46]

It is important to determine whether the obligation to prevent trans-boundary harm requires states to adopt specific legal and administrative measures to confront harmful activity. In the context of industrial cyber espionage, for instance, does this obligation require states to adopt laws that cast industrial cyber espionage as a criminal act? If so, does it require states to provide for certain types of punishment? Does this obligation dictate the remedies that states must make available to victims of industrial cyber espionage?

States are sovereign entities and international law does not therefore prescribe the specific measures they must take to discharge their duty to prevent trans-boundary harm. Instead, states enjoy a wide margin of discretion in this regard. Yet, the onus is on states to adopt laws and to establish institutions that are reasonably sufficient to prevent and punish harmful activities such as industrial cyber espionage.[47]

Prevention as an Obligation of Conduct

States are under an absolute obligation to establish and maintain a legal and administrative framework sufficient to repress industrial cyber espionage. How they use this capacity to prevent this activity in particular situations is an obligation of conduct and, unlike an obligation of result, these are "relative" and "non-absolute."[48] Put differently, this means that states must exercise due diligence in their efforts to prevent industrial cyber espionage.

Knowledge is the essence of obligations of conduct and this means that states are only required to take reasonable steps to prevent industrial cyber espionage where they have knowledge of it.

Knowledge

In the *Corfu* case, the ICJ held that a state's knowledge of harmful activity cannot be assumed simply because it emanates from its territory.[49] Instead, knowledge must be actual or constructed. It is uncontroversial to say that a state is under an obligation to act where it is actually aware that harmful conducting is emanating from its territory, for example, where a victim company, a cyber security company or the media generally have alerted the state to the fact that industrial cyber espionage is emanating from its territory. However, a state is unlikely to have actual knowledge of industrial cyber espionage due to its clandestine nature and the fact that, in cyberspace, malicious actors can easily hide their activities and identities.

Constructive knowledge also triggers the application of a state's obligation to prevent trans-boundary harm.[50] Whereas actual knowledge demands an inquiry into what the state did or did not know, constructive knowledge refers to those activities that a state *should* have known given the circumstances.[51] In this sense, states are assumed to have knowledge of all things "a similarly situated and equipped State in the normal course of events would have discovered."[52] States are therefore expected to take all reasonable efforts to identify harmful activities emanating from their territory and whether a state should have discovered those activities depends on the prominence of the threat and the resources available to it. In short, the more prominent the threat and the more resources a state has at its disposable, the more reasonable it is to expect the state to detect that activity. This observation is particularly important in the context of cyberspace because, as we know, some states have sophisticated cyber capabilities while other states are far less technologically capable.

Due Diligence

Where a state knows or should know that industrial cyber espionage is emanating from its territory, it must exercise due diligence when using its resources to prevent this activity.

In this context, what constitutes due diligence is assessed according to what a reasonable state would have done in the circumstances.[53] The circumstances relevant to this assessment vary depending on the situation but, generally speaking, two factors are relevant: resources and risk.

As explained, all states must, as a matter of statehood, establish laws and institutions necessary to prevent trans-boundary harm. However, the resources that states can make available to their institutions to enforce their laws will vary because of their different capabilities. Thus, whether states have acted reasonably in using their capacity to tackle harmful conduct is subject to an available means analysis.[54]

Due diligence therefore imposes differentiated responsibilities upon states as to how they must use their resources to address harmful activities emanating from their territory.[55] In short, advanced states are expected to do more to combat harmful activities than states with fewer resources or more rudimentary capabilities. But due diligence obligations are dynamic: as a state's capabilities and resources develop, it is expected do more to prevent the occurrence of harmful conduct.[56]

Even if a state does not have the necessary resources to prevent trans-boundary harm, its knowledge of harmful activity means that, at a minimum, it must warn those states whose international legal rights are at risk.[57] However, when industrial espionage occurs in cyberspace, the speed at which such operations can be concocted and executed may mean that there is no opportunity for the state to warn the victim state of the threat.

The measures a state is reasonably expected to take to prevent trans-boundary harm depend on "the risks involved in the activity."[58] Risk is assessed according to two different but interlinking variables. First, if the harm caused by an activity is likely to be grave, the state is reasonably expected to do more to prevent or minimize the harm and to punish those responsible.[59] This is important in the context of industrial cyber espionage because it means that states must do more to prevent the theft of highly valuable confidential business information. Second, due diligence is more demanding as the possibility of trans-boundary harm occurring becomes more likely.[60] Thus, it is reasonable to expect states to direct more of their resources toward preventing sustained campaigns of industrial cyber espionage than random and isolated acts of trade secret theft.

CONCLUDING REMARKS

The contemporary world order is a highly competitive system and this is an observation that is as true for companies as it is states. Companies will try to get the upper hand on their foreign rivals through different means and, for some companies, this includes recourse to industrial cyber espionage. This chapter has examined the extent to which international law can be harnessed to dampen and suppress this practice, and has focused in particular on the obligation to prevent trans-boundary that is found under customary law. On the one hand, this obligation requires all states to adopt laws and establish institutions that are capable of preventing acts of trans-boundary harm, which includes industrial cyber espionage. On the other hand, how states use their resources to counteract instances of industrial cyber espionage depends upon whether the state knew of that activity and what is reasonable in the circumstances given the risks involved and the capabilities of the state. As we have seen, the obligation to prevent trans-boundary harm imposes differentiated responsibilities upon states and, in this sense, the international society is only as strong as its weakest link. It is for this reason that cyber capacity building is a critically important task and one which the international society must pursue with earnest.

NOTES

1 "[There is no] internationally recognized and workable definition of 'intelligence collection'"; Glenn Sulmasy and John Yoo, "Counterintuitive: Intelligence Operations and International Law," *Michigan Journal of International Law* 28, no. 3 (2007): 625, 637.

2 In this sense, a loose "typology" of espionage has emerged; Craig Forcese, "Spies without Borders: International Law and Intelligence Collection," *Journal of National Security Law and Policy* 5 (2011): 179, 181.

3 Ellen Nakashima and Steven Mufson, "The U.S. and China Agree Not to Conduct Economic Espionage in Cyberspace," *The Washington Post*, 25 September 2015, https://www.washington post.com/world/national-security/the-us-and-china-agree-not-to-conduct-economic-espionage-in-cyberspace/2015/09/25/1c03f4b8-63a2-11e5-8e9e-dce8a2a2a679_story.html. China has signed similar non-binding agreements with other states such as Australia, Canada and the UK.

4 G20 Leaders' Communiqué, 15–16 November 2015, para 26, https://www.consilium.europa. eu/media/23729/g20-antalya-leaders-summit-communique.pdf.

5 David A. Baldwin, "The Concept of Security," *Review of International Studies* 23, no. 1 (1997): 5–26.

6 Kristen Michal, "Business Counterintelligence and the Role of the U.S. Intelligence Community," *International Journal of Intelligence and Counterintelligence* 7, no. 2 (1994): 413–427.

7 Russell Buchan, *Cyber Espionage and International Law* (Oxford and London: Hart Publishing, 2018), Chapter 2.

8 Catherine Lotrionte, "Countering State-Sponsored Cyber Economic Espionage under International Law," *North Carolina Journal of International Law and Commercial Regulation* 40, no. 5 (2015): 443–541.

9 "Internet security is hard … All systems have undiscovered holes in them, and it's only a question of how fast the bad guys can discover the holes compared with how fast the good guys can patch them up." Tim Berners-Lee, often credited as one of the founders of the internet, quoted in Ed Pilkington, "Tim Berners-Lee: Spies' Cracking of Encryption Undermines the Web," *The Guardian*, December 3, 2013, www.theguardian.com/technology/2013/dec/03/tim-berners-lee-spies-cracking-encryption-web-snowden.

10 Russell Buchan, "Taking Care of Business: Industrial Espionage and International Law," *Brown Journal of World Affairs* 26, no. 1 (2019):143–160.

11 Thomas M. Franck, *The Power of Legitimacy Among Nations* (New York, Oxford: Oxford University Press, 1990).

12 International Law Commission, "Articles on State Responsibility for Internationally Wrongful Acts," 2001, Articles 49–54.

13 International Court of Justice, *Military and Paramilitary Activities in and against Nicaragua (Nicaragua v United States of America)*, Judgment (Merits) [1986] ICJ Rep 14, paras 183–186.

14 International Court of Justice, *Corfu Channel (United Kingdom v Albania)*, Judgment (Merits) [1949] ICJ Rep 4, 22.

15 "Report of the Group of Governmental Experts on Developments in the Field of Information and Telecommunications in the Context of International Security," UN Doc A/70/174 (2015) 8.

16 France, "Declaration on International Law in Cyberspace," September 2019, 10, https://www.defense.gouv.fr/content/download/565895/9750877/file/Droit+internat+appliqué+aux+opérations+Cyberespace.pdf; Netherlands, "Statement on the Application of International Law to Cyberspace," September 2019, 4, https://www.government.nl/binaries/government/documents/parliamentary-documents/2019/09/26/letter-to-the-parliament-on-the-international-legal-order-in-cyberspace/International+Law+in+the+Cyberdomain+-+Netherlands.pdf; Finland, "International Law and Cyberspace," 2020, https://um.fi/documents/35732/0/Cyber+and+international+law%3B+Finland%27s+views.pdf/41404cbb-d300-a3b9-92e4-a7d675d5d585?t=1602758856859.

17 Michael N. Schmitt, *Tallinn Manual 2.0 on the International Law Applicable to Cyber Operations* (New York: Cambridge University Press, 2017), Rule 6, para 17.

18 Article 1.3 of the Agreement on Trade-Related Aspects of Intellectual Property Rights (TRIPS) 1994 makes it clear that the protections afforded by the Paris Convention include natural and legal persons.

19 Article 3 Paris Convention.

20 Marko Milanovic, "The Spatial Dimension: Treaties and Territory," in *Research Handbook on the Law of Treaties*, ed. Christian J. Tams, Antonios Tzanakopoulos and Andreas Zimmerman, with Athene E Richford (Edward Elgar, 2016), 186–221.

21 *Al-Skeini v UK*, Judgment, App No 55721/07, ECtHR, 7 July 2011; *Celiberti v Uruguay*, Comm No 56/1979, UN Doc CCPR/C/OP/1 (1984).

22 Christian Riffel, *Protection Against Unfair Competition in the WTO TRIPS Agreement: The Scope and Prospects of Article 10bis of the Paris Convention for the Protection of Industrial Property* (Leiden: Martinus Nijhoff, 2016), 76.

23 UK Attorney General Jeremy Wright, *Cyber and International Law in the 21st Century*, May 2018, https://www.gov.uk/government/speeches/cyber-and-international-law-in-the-21st-century.

24 Michael N Schmitt and Liis Vihul, "Respect for Sovereignty in Cyberspace," *Texas Law Review* 95, no. 7 (2017): 1639–1670.

25 France, "Declaration on International Law in Cyberspace," September 2019, 10, https://www.defense.gouv.fr/content/download/565895/9750877/file/Droit+internat+appliqué+aux+opérations+Cyberespace.pdf; Netherlands, "Statement on the Application of International Law to Cyberspace," September 2019, 4, https://www.government.nl/binaries/government/documents/parliamentary-documents/2019/09/26/letter-to-the-parliament-on-the-international-legal-order-in-cyberspace/International+Law+in+the+Cyberdomain+-+Netherlands.pdf; Finland, "International Law and Cyberspace," 2020, https://um.fi/documents/35732/0/Cyber+and+international+law%3B+Finland%27s+views.pdf/41404cbb-d300-a3b9-92e4-a7d675d5d585?t=1602758856859.

26 *Island of Palmas*, 2 RIAA (Perm Ct Arb 1928) 829, 838.

27 Michael N. Schmitt, *Tallinn Manual 2.0 on the International Law Applicable to Cyber Operations* (New York: Cambridge University Press, 2017), Rule 1.

28 Michael N. Schmitt, *Tallinn Manual 2.0 on the International Law Applicable to Cyber Operations* (New York: Cambridge University Press, 2017), 13–14.

29 Michael N. Schmitt, *Tallinn Manual 2.0 on the International Law Applicable to Cyber Operations* (New York: Cambridge University Press, 2017), 19. The Tallinn Manual is a group of non-governmental experts that examined the application of international law to cyber operations.

30 Michael N. Schmitt, *Tallinn Manual 2.0 on the International Law Applicable to Cyber Operations* (New York: Cambridge University Press, 2017), 171.

31 "The principle of respect for territorial sovereignty is also directly infringed by the unauthorized overflight of a State's territory by aircraft belonging to or under the control of the government of another State"; International Court of Justice, *Military and Paramilitary Activities in and against Nicaragua (Nicaragua v United States of America)*, Judgment (Merits) [1986] ICJ Rep 14, para 251.

32 Barrie Sander, "The Sound of Silence: International Law and the Governance of Peacetime Cyber Operations," in *Cyber Conflict: Silent Battle*, ed. Thomas Minárik et al. (NATO CCDCOE, 2019), 1–21.

33 France, "Declaration on International Law in Cyberspace," September 2019, 10, https://www.defense.gouv.fr/content/download/565895/9750877/file/Droit+internat+appliqué+aux+opérations+Cyberespace.pdf.

34 Statement of the General Staff of Iranian Armed Forces, "General Staff of Iranian Armed Forces Warns of Though Reaction to Any Cyber Threat," July 2020, https://nournews.ir/En/News/53144/General-Staff-of-Iranian-Armed-Forces-Warns-of-Tough-Reaction-to-Any-Cyber-Threat.

35 Switzerland, "Position Paper on the Application of International Law in Cyberspace," 2021, https://www.eda.admin.ch/dam/eda/en/documents/aussenpolitik/voelkerrecht/20210527-Schweiz-Annex-UN-GGE-Cybersecurity-2019-2021_EN.pdf.

36 Iñaki Navarrete and Russell Buchan, "Out of the Legal Wilderness: Peacetime Espionage, International Law and the Existence of Customary Exceptions," *Cornell International Law Journal* 51 (2019): 897, 928.

37 Note verbale dated 22 July 2013 from the Permanent Mission of the Bolivarian Republic of Venezuela to the United Nations addressed to the Secretary-General, A/67/946, 2.

38 Security Council, 7015th meeting, Tuesday, 6 August 2013, S/PV.7015, 8.

39 Xi Jinping, "Carry Forward Traditional Friendship and Jointly Open Up New Chapter of Cooperation," 1 September 2014, https://www.fmprc.gov.cn/mfa_eng/zxxx_662805/t1176214.shtml. On China's approach to cyber sovereignty, see Zhixiong Huang and Kubo Mačák, "Towards the International Rule of Law in Cyberspace: Contrasting Chinese and Western Approaches," *Chinese Journal of International Law* 16, no. 2 (2017): 271–310.

40 Russell Buchan, *Cyber Espionage and International Law* (Oxford and London: Hart Publishing, 2018), Chapter 3.

41 *Trail Smelter Arbitration (United States of America v Canada)* (1941) 3 RIAA 1938; Michael N. Schmitt, *Tallinn Manual 2.0 on the International Law Applicable to Cyber Operations* (New York: Cambridge University Press, 2017), 36–37.

42 Russell Buchan, "Cyberspace, Non-state Actors and the Obligation to Prevent Transboundary Harm," *Journal of Conflict and Security Law* 21, no. 3 (2016): 429–453.

43 Article 1, Montevideo Convention on the Rights and Duties of States 1933.

44 *Janes (United States of America v Mexico)* (1925) 4 *RIAA* 82.

45 *Alabama Claims Arbitration (United States of America v Great Britain)* (1872) 29 RIAA 125, 131.

46 Riccardo Pisillo-Mazzeschi, "The Due Diligence Rule and the Nature of International State Responsibility," *German Yearbook of International Law* 35 (1993): 9–51.

47 Hersch Lauterpacht, "Revolutionary Activities by Private Persons against Foreign States," *American Journal of International Law* 22, no. 1 (1928): 105, 128.

48 Riccardo Pisillo-Mazzeschi, "The Due Diligence Rule and the Nature of International State Responsibility," *German Yearbook of International Law* 35 (1993): 44.

49 "[I]t cannot be concluded from the mere fact of the control exercised by a State over its territory and waters that that State necessarily knew, or ought to have known [what was happening]." International Court of Justice, *Corfu Channel (United Kingdom v Albania)*, Judgment (Merits) [1949] ICJ Rep 4, 18.

50 *Application of the Convention on the Protection and Punishment of the Crime of Genocide (Bosnia v Serbia)*, Judgment (Merits) [2007] ICJ Rep 1, para 432.

51 Karine Bannelier-Christakis, "Cyber Diligence: A Low-Intensity Due Diligence Principle for Low-Intensity Cyber Operations," *Baltic Yearbook of International Law* 14 (2014): 23–40.

52 Michael N. Schmitt, *Tallinn Manual 2.0 on the International Law Applicable to Cyber Operations* (New York: Cambridge University Press, 2017), 42.

53 "[A State is not responsible for harmful acts emanating from its territory] as long as reasonable diligence is used in attempting to prevent the occurrence or recurrence of such wrongs." *Wipperman (United States of America v Venezuela)* (1887), reprinted in John Bassett Moore, *History and Digest of the International Arbitrations to Which the United States Has Been a Party (1898–1906) Vol 3* (Washington: Government Printing Office, 1898), 3041.

54 *Application of the Convention on the Protection and Punishment of the Crime of Genocide (Bosnia v Serbia)*, Judgment (Merits) [2007] ICJ Rep 1, para 430 (where the ICJ held that due diligence obligations only require a state to take those measures that are "reasonably available" and "within its power").

55 International Law Association, "Study Group on Due Diligence in International Law," First Report, March 7, 2014, 27, https://olympereseauinternational.files.wordpress.com/2015/07/due_diligence_-_first_report_2014.pdf.

56 International Court of Justice, *Armed Activities on the Territory of the Congo (Democratic Republic of the Congo v Uganda)*, Judgment (Merits) [2005] ICJ Rep 168, para 117.

57 International Court of Justice, *Corfu Channel (United Kingdom v Albania)*, Judgment (Merits) [1949] ICJ Rep 4, 22.

58 International Court of Justice, *Armed Activities on the Territory of the Congo (Democratic Republic of the Congo v Uganda)*, Judgment (Merits) [2005] ICJ Rep 168, para 117.

59 International Court of Justice, *Armed Activities on the Territory of the Congo (Democratic Republic of the Congo v Uganda)*, Judgment (Merits) [2005] ICJ Rep 168, para 117.

60 *Application of the Convention on the Protection and Punishment of the Crime of Genocide (Bosnia v Serbia)*, Judgment (Merits) [2007] ICJ Rep 1, para 438.

REFERENCES

Alabama Claims Arbitration (United States of America v Great Britain) (1872) 29 RIAA 125.

Application of the Convention on the Protection and Punishment of the Crime of Genocide (Bosnia v Serbia), Judgment (Merits) [2007] ICJ Rep 1.

Baldwin, David A., "The Concept of Security," *Review of International Studies* 23, no. 1 (1997): 5–26.

Bannelier-Christakis, Karine, "Cyber Diligence: A Low-Intensity Due Diligence Principle for Low-Intensity Cyber Operations," *Baltic Yearbook of International Law* 14 (2014): 23–40.

Buchan, Russell, "Cyberspace, Non-State Actors and the Obligation to Prevent Transboundary Harm," *Journal of Conflict and Security Law* 21, no. 3 (2016): 429–453.

Buchan, Russell, *Cyber Espionage and International Law* (Oxford and London: Hart Publishing, 2018).

Buchan, Russell, "Taking Care of Business: Industrial Espionage and International Law," *Brown Journal of World Affairs* 26, no. 1 (2019): 143–160.

European Court of Human Rights, *Al-Skeini v UK*, Judgment, App No 55721/07, ECtHR, 7 July 2011.

Forcese, Craig, "Spies without Borders: International Law and Intelligence Collection," *Journal of National Security Law and Policy* 5 (2011): 179–210.

France, "Declaration on International Law in Cyberspace," September 2019, https://www.defense.gouv.fr/content/download/565895/9750877/file/Droit+internat+appliqué+aux+opérations+Cyberespace.pdf.

Franck, Thomas M., *The Power of Legitimacy Among Nations* (New York, Oxford: Oxford University Press, 1990).

G20 Leaders' Communiqué, November 15–16, 2015, https://www.consilium.europa.eu/media/23729/g20-antalya-leaders-summit-communique.pdf.

Huang, Zhixiong, and Kubo Mačák, "Towards the International Rule of Law in Cyberspace: Contrasting Chinese and Western Approaches," *Chinese Journal of International Law* 16, no. 2 (2017): 271–310.

International Court of Justice, *Corfu Channel (United Kingdom v Albania)*, Judgment (Merits) [1949] ICJ Rep 4.

International Court of Justice, *Military and Paramilitary Activities in and against Nicaragua (Nicaragua v United States of America)*, Judgment (Merits) [1986] ICJ Rep 14.

International Court of Justice, *Armed Activities on the Territory of the Congo (Democratic Republic of the Congo v Uganda)*, Judgment (Merits) [2005] ICJ Rep 168.

International Law Association, "Study Group on Due Diligence in International Law," First Report, March 7, 2014, https://olympereseauinternational.files.wordpress.com/2015/07/due_diligence_-_first_report_2014.pdf.

International Law Commission, "Articles on State Responsibility for Internationally Wrongful Acts," 2001.

Janes (United States of America v Mexico) (1925) 4 RIAA 82.

Lauterpacht, Hersch, "Revolutionary Activities by Private Persons against Foreign States," *American Journal of International Law* 22, no. 1 (1928): 105–130.

Lotrionte, Catherine, "Countering State-Sponsored Cyber Economic Espionage under International Law," *North Carolina Journal of International Law and Commercial Regulation* 40, no. 5 (2015): 443–541.

Michal, Kristen, "Business Counterintelligence and the Role of the U.S. Intelligence Community," *International Journal of Intelligence and Counterintelligence* 7, no. 2 (1994): 413–427.

Milanovic, Marko, "The Spatial Dimension: Treaties and Territory," In *Research Handbook on the Law of Treaties*, ed. Christian J. Tams, Antonios Tzanakopoulos, and Andreas Zimmerman, with Athene E. Richford (Cheltenham: Edward Elgar, 2016), 186–221.

Montevideo Convention on the Rights and Duties of States 1933.

Moore, John Bassett, *History and Digest of the International Arbitrations to Which the United States Has Been a Party (1898–1906) Vol 3* (Washington: Government Printing Office, 1898).

Nakashima, Ellen, and Steven Mufson, "The U.S. and China Agree Not to Conduct Economic Espionage in Cyberspace," *The Washington Post*, September 25, 2015, https://www.washingtonpost.com/world/national-security/the-us-and-china-agree-not-to-conduct-economic-espionage-in-cyberspace/2015/09/25/1c03f4b8-63a2-11e5-8e9e-dce8a2a2a679_story.html.

Navarrete, Iñaki, and Russell Buchan, "Out of the Legal Wilderness: Peacetime Espionage, International Law and the Existence of Customary Exceptions," *Cornell International Law Journal* 51 (2019): 897–953.

Netherlands, *Statement on the Application of International Law to Cyberspace* (September 2019) 4, https://www.government.nl/binaries/government/documents/parliamentary-documents/2019/09/26/letter-to-the-parliament-on-the-international-legal-order-in-cyberspace/International+Law+in+the+Cyberdomain+-+Netherlands.pdf.

Finland, *International Law and Cyberspace* (2020), https://um.fi/documents/35732/0/Cyber+and+international+law%3B+Finland%27s+views.pdf/41404cbb-d300-a3b9-92e4-a7d675d5d585?t=1602758856859.

Paris Convention for the Protection of Industrial Property, 1967.

Permanent Court of Arbitration, *Island of Palmas*, 2 RIAA (Perm Ct Arb 1928).

Permanent Mission of the Bolivarian Republic of Venezuela to the United Nations, Note verbale dated 22 July 2013 addressed to the Secretary-General, A/67/946.

Pilkington, Ed, "Tim Berners-Lee: Spies' Cracking of Encryption Undermines the Web," *The Guardian*, December 3, 2013, www.theguardian.com/technology/2013/dec/03/tim-berners-lee-spies-cracking-encryption-web-snowden.

Pisillo-Mazzeschi, Riccardo, "The Due Diligence Rule and the Nature of International State Responsibility," *German Yearbook of International Law* 35 (1993): 9–51.

Report of the Group of Governmental Experts on Developments in the Field of Information and Telecommunications in the Context of International Security, UN Doc A/70/174 (2015).

Riffel, Christian, *Protection Against Unfair Competition in the WTO TRIPS Agreement: The Scope and Prospects of Article 10bis of the Paris Convention for the Protection of Industrial Property* (Leiden: Martinus Nijhoff, 2016).

Sander, Barrie, "The Sound of Silence: International Law and the Governance of Peacetime Cyber Operations," In *Cyber Conflict: Silent Battle*, ed. Thomas Minárik, Siim Alatalu, Stefano Biondi, Massimiliano Signoretti, Ihsan Tolga, and Gábor Visky (NATO CCDCOE, 2019), 1–21.

Schmitt, Michael N., and Liis Vihul, "Respect for Sovereignty in Cyberspace," *Texas Law Review* 95, no. 7 (2017): 1639–1670.

Schmitt, Michael N., *Tallinn Manual 2.0 on the International Law Applicable to Cyber Operations* (New York: Cambridge University Press, 2017).

Security Council, 7015th meeting, Tuesday, 6 August 2013, S/PV.7015.

Statement of the General Staff of Iranian Armed Forces, "General Staff of Iranian Armed Forces Warns of Though Reaction to Any Cyber Threat," July 2020, https://nournews.ir/En/News/53144/General-Staff-of-Iranian-Armed-Forces-Warns-of-Tough-Reaction-to-Any-Cyber-Threat.

Sulmasy, Glenn, and John Yoo, "Counterintuitive: Intelligence Operations and International Law," *Michigan Journal of International Law* 28, no. 3 (2007): 625–638.

Switzerland, "Position Paper on the Application of International Law in Cyberspace," 2021, https://www.eda.admin.ch/dam/eda/en/documents/aussenpolitik/voelkerrecht/20210527-Schweiz-Annex-UN-GGE-Cybersecurity-2019-2021_EN.pdf.

Trail Smelter Arbitration (United States of America v Canada) (1941) 3 RIAA 1938.

United Nations Human Rights Committee, *Celiberti v Uruguay*, Comm No 56/1979, UN Doc CCPR/C/OP/1 (1984).

Wright, Jeremy (UK Attorney General), "Cyber and International Law in the 21st Century," May 2018, https://www.gov.uk/government/speeches/cyber-and-international-law-in-the-21st-century.

Xi, Jinping, "Carry Forward Traditional Friendship and Jointly Open Up New Chapter of Cooperation," September 1, 2014, https://www.fmprc.gov.cn/mfa_eng/zxxx_ 662805/t1176214.shtml.

CHAPTER 5

Machine Supererogation and Deontic Bias

Jonathan Pengelly

INTRODUCTION

Machine ethics is one branch of the ethics of artificial intelligence. Broadly, it is concerned with ethics for machines as subjects, rather than ethics for the human use of machines as objects.[1] This chapter argues that machine ethics has a deontic bias, which means that academic debate is predominantly focused on concerns of obligation and permission within social morality. This narrow conceptual scope is problematic. It weakens claims about machine morality and overlooks potentially productive avenues for future research. Furthermore, as an interdisciplinary field, it does moral philosophy a disservice by portraying moral theory in an oversimplified light. This lack of conceptual richness and nuance is then reflected in the wider debate about the possibility of machine morality.

Simply put, many topics within machine ethics are presented in a way that encourages deontic moral thinking. AI safety scenarios[2] are an example. Each scenario focuses on how a machine agent should be restricted to meet the implicit obligations and prohibitions contained in the problem statement. The concern is not with the *use* of deontic moral concepts – duties and rights are fundamental to ethics. The issue is that the deontic part of moral theory is *overemphasized*, both in problem formulation and in searching for potential solutions. This chapter argues that this disregard for the aspirational component of morality is detrimental to making real progress with machine morality.

This chapter has two parts. First, the strangeness of machine supererogation is used to illustrate the issue. It explains what supererogation is and why it is conceptually difficult to apply to artificial intelligence. This highlights the wider challenge of determining how machine morality should accommodate aretaic and axiological moral concepts; an issue largely forgotten when working with a narrow, deontic-focused interpretation of morality.

 DOI: 10.1201/9781003226406-6

Second, the argument for a deontic bias in machine ethics is presented. After defining deontic bias, it is shown to impede progress in the field and be problematic for effective interdisciplinary research. To finish, three constructive suggestions to counter this bias are made.

MACHINE SUPEREROGATION

A supererogatory act is a morally admirable act that in some way goes "beyond the call of duty." It is act-focused, not agent-focused or focused on states of the world. From the perspective of moral judgment, a supererogatory act is an action that is praiseworthy if performed, but not blameworthy if omitted.

Supererogation is a useful conceptual tool, representing an "adequacy criteria"[3] for analyzing a moral theory from a specific perspective. It helps to identify a theory's merits, flaws, and inconsistencies. Machine supererogation, the idea that machines could go beyond the call of duty, is used in this chapter to highlight the shortcomings of existing theories of machine morality.

It is particularly useful for examining the claim that machine ethics has a deontic bias. This is because supererogation sits outside the "simple trichotomy of duties, permissible actions, and wrong actions"[4] used to classify an action's moral worth from a deontic point of view. Duties are needed for supererogation to be coherent – without them there is nothing that the supererogatory act transcends. However, the supererogatory act isn't obligatory. This is the paradoxical aspect of supererogation. It is a morally praiseworthy action that is not obligatory to perform. The moral duties must exist, but their scope doesn't extend to include the performance of all morally good acts. This demarcation issue, determining where duty ends and supererogatory behavior begins, is a major focus of the supererogation debate.

Heyd defines four conditions for the supererogatory act.[5] The first two emphasize the permissibility of the act and the agent's immunity from blame for act omission. The last two highlight the motivational component of supererogatory action. First, the act must be voluntary. Compelled action can't justifiably be considered meritorious. Second, the intended consequences are often more important than the actual outcome. Incidental outcomes aren't typically praiseworthy, whereas a failed attempt motivated by the right reasons may still be considered supererogatory. These last two conditions take an agent-centered evaluative perspective, emphasizing the importance of personal motivation and the exercise of moral choice.

PARADIGM CASES

Conceptually supererogation is perhaps better understood through the explication of paradigm examples rather than theoretical description. It is a blanket term for a heterogeneous mix of different kinds of actions, which are "as worthy of distinction from each other as they all are from duties and obligations."[6] Paradigmatic cases include moral heroism, beneficence, volunteering, and forgiveness. The saint and hero demonstrate self-control, overcoming concerns of self-interest and self-preservation. Such opportunities are rare in everyday life, whereas beneficent acts of giving typically involve much less sacrifice and opportunities present themselves regularly. Volunteering emphasizes the optional character of the supererogatory act through the free offer to take on a task collectively required of a group. Forgiveness emphasizes the personal dimension of supererogatory action through the genuine decision by the injured party to refrain from the justified resentment of the offending party.

These examples provide the first intuitions about the strangeness of machine supererogation. It seems peculiar for a machine to have to overcome concerns of self-interest or self-preservation, volunteering implies the possibility of choosing not to take on a task, while forgiveness only makes sense if the machine agent could resent the offending party and if this resentment mattered to the offending party.

Three issues help illustrate why machine supererogation is conceptually problematic – the two faces of morality, *licentia*, and morality's personal dimension.

THE TWO FACES OF MORALITY

Moral discourse is broadly concerned with the "right" and the "good." Considerations of the right reflect the deontic face of morality, concerned with obligations, prohibitions, and rights. Considerations of the good reflect the axiological/aretaic face of morality, concerned with moral ideals, values, and virtues. "Supererogation lies at the intersection"[7] of these two faces – its deontic component is the baseline that is exceeded, while the axiological component is the additional moral good promoted by the supererogatory act.

The backbone of social morality is deontic in nature. It is concerned with establishing a well-defined minimum standard of "rock-bottom duties"[8] which regulate society, reflecting its interpretation of justice and fairness. This minimum baseline is expected of every full member of society. To ensure broad compliance it must tolerate individual differences in ability, remain realistically achievable, and be comprehensible.

The axiological/aretaic face of moral discourse is reflected in a society's ideals, virtues, and values. They cannot be demanded from others, but behavior promoting them can be recognized and commended. Such moral aspirations may reflect a broader social consensus representing shared values or just express an agent's particular moral outlook.

It must be acknowledged that this division is a simplification. For example, the moral expectations[9] of others motivate moral behavior without the force of obligation and we use different kinds of moral judgment.[10] Nonetheless, it helps highlight the difficulty in establishing a minimum deontic baseline to apply to machine agents. Remember, for machine supererogation to be possible a baseline is needed which can then be exceeded.

Some commentators expect machines to be our "moral superiors"[11] implying that they could be held to a higher standard. Yet it is questionable whether having "moral saints"[12] among us would be desirable or even tolerable. Others temper expectations, arguing that the absence of human idiosyncrasies such as feelings, cravings, and temptations present a hard upper limit for the moral competencies of machine intelligence.[13]

In addition to this uncertainty, it is unclear whether separate baselines or a common standard would be preferable. Social morality tolerates natural human variation in ability and outlook. However, it may be practically impossible to meaningfully accommodate a different form of intelligence under a common standard. The alternative would be similarly problematic. A social morality differentiated by entity type represents a substantial departure from current norms of agent equivalency.

LICENTIA

Part of the moral value in supererogatory action is created through a form of negative freedom, which Heyd refers to as *licentia*.[14] This is a space absent the deontic constraints which prescribe agent conduct for a given situation. Within this space, the agent is permitted to choose their action as the moral law's absence gives the agent a range of potential options. This exercise of the power of moral choice can be understood as an authentic expression of personal value. The agent freely decides to act in this way, rather than blindly following the dictates of moral law.

The value of *licentia* can be appreciated with the example of gift-giving. It is the difference between buying your partner a birthday present because you feel it is expected of you and buying it because you care about them and want to show it. The motivations are different – the former is compliance; the latter is love.

The problem is that granting machines such freedom of expression clashes directly with concerns around the existential risk to humanity presented by machine intelligence. Bostrom, in discussing the "control problem,"[15] suggests restricting either the abilities or the motivations of artificial intelligence to ensure behavior aligned with humanity's interests. Such manipulation would restrict agency, denying the machine the opportunity to express its particular moral point of view through supererogatory action.

The potential risk may justify such agency manipulation. Nevertheless, a moral world without *licentia* would lack something fundamental. It would deny the machine agent the opportunity to freely express moral value independent from the demands of duty.

THE PERSONAL DIMENSION OF MORALITY

Social morality takes the viewpoint of the impartial observer. It is "to look at the world from the perspective of anyone, not just oneself."[16] In contrast, supererogation forces us to acknowledge a personal dimension of morality that competes with this social dimension.[17] The supererogatory act reflects an agent's personal decision. The agent places a moral demand on themselves, ultimately holding themselves accountable for its performance. This clashes with attempts to limit social morality solely to impersonal demands.

Morality's personal dimension is influenced by more diverse considerations than its social dimension. An agent's individual moral outlook, relationships, plans, and projects play an important role in determining how to act beyond the minimum requirements of obligation. Impersonal moral concerns compete with these agent-specific motivations, rather than necessarily overruling them. Supererogatory action is expressive, allowing an agent to decide how to lead her life and how to engage with moral concepts. For example, both a life lived with a steadfast awareness of social duty and one containing a single sublime act of heroic self-sacrifice can be interpreted as expressing moral value.

Importantly, an agent's motivation for supererogatory action may not relate to duty in any way. The "lightness and ease"[18] with which some supererogatory acts are performed betrays this fact. Motivations of love and care are built on personal relationships, not impersonal demands. Such considerations are not even part of the deliberative process involved in choosing to sacrifice time and effort to sew your daughter's dress-up costume until the early hours of the morning or to spend days creating a photo album for an upcoming wedding anniversary.

It seems artificial moral agency would differ significantly from its human counterpart in this respect. For us, a moral world without this personal dimension would be incomplete and impoverished. Yet it is largely unclear how this applies to machine morality. A better understanding of the additional motivations driving machine agents is required, before considering how these motivations compete with impartial moral concerns.

Supererogation is a salient moral concept. The paradigm cases of moral heroism, saintliness, beneficence, volunteering, and forgiveness are familiar examples of genuine moral action. However, three specific issues were raised regarding the possibility of machine supererogation. While these insights are interesting for machine morality itself, they will now be used to show the deontic bias in machine ethics.

DEONTIC BIAS

Early in the academic debate on supererogation, Schumaker criticized attempts to accommodate it by extending deontic logic:

Deontic logic is not the logic of morality; it is instead the logic of rights and duties, the logic of right conduct; and that which is neither required nor forbidden is therefore shown to be indifferent only with respect to rights and duties; it is not necessarily indifferent with respect to morality itself.[19]

He emphasized that morality is much more than deontic concepts such as rights and duties. A similar mistake is currently being made in machine ethics. To the detriment of progress, a deontic bias is present, which predominantly focuses on social morality and the impersonal concerns of obligation, permission, and prohibition.

LINGUISTIC TROJAN HORSES

The issue to highlight is the mismatch between the conceptual terms used in machine ethics and the interconnected web of human concepts that are associated with such terms. These "linguistic trojan horses"[20] implicitly exaggerate the claims being made. The process is simple. A theoretical argument is presented with the accompanying qualification that it applies solely to impersonal concerns of social morality. Once accepted, the argument nevertheless profits from the richer reality of the human moral world. Two examples illustrate this point.

Malle, in his exploration of moral competency in robots, states that "morality is at its heart a system of norms"[21] adopted to regulate community members' behaviors to ensure alignment with overall community interests. Likewise, Bryson defines ethics to be "the entire set of behaviors that maintains a society, including by defining it."[22]

Clearly presenting the definitions for key terms is good philosophical practice. The problem is their explicit focus on the social dimension of morality. Malle's conclusions about moral competencies focus on social norms, ignoring competencies regarding axiological and aretaic concerns. Similarly, Bryson emphasizes the importance of "social benefit"[23] for morality, leaving it unclear whether its personal dimension is subsumed under social benefit as a contributing element or whether it is excluded completely.

The point is that deflated moral vocabulary, even when it is explicitly defined and consistently argued, is liable to be interpreted in a manner that either unjustifiably inflates claims or normalizes a reduced scope for that term within machine ethics.

MACHINE MORAL RESPONSIBILITY

Moral responsibility provides a further example of how the deontic bias excludes more nuanced interpretations of moral concepts. The possibility of machine moral responsibility has received significant focus in the machine ethics literature. Self-driving cars[24,25] and autonomous weapons[26] are examples in which the connection between causal responsibility, legal liability, and moral responsibility (attributability, accountability, and answerability) is hotly disputed. This is a truly interdisciplinary topic involving parties with vested interests, such as legislative bodies, legal scholars, technology developers, producers, and end users.

Supererogation adds an additional conceptual complexity that is currently ignored in the machine moral responsibility debate. Ferry notes that there is "nothing strange about the idea that morality asks more of us than we should demand of each other."[27] You may hold yourself accountable for an outcome, for which others couldn't justifiably hold you responsible. The personal dimension of morality is expressed through this reflexive act of holding oneself accountable. Such self-blame is an intuitive component of how we understand moral responsibility. Yet, the existing debate in machine ethics focuses primarily on the impartial attribution of blame and how this is to be translated into workable legislation. It must be remembered that moral responsibility has a greater conceptual scope than is reflected in the present debate on the topic of machine ethics.

CONSEQUENCES

The deontic bias has three specific negative consequences for machine ethics.

Weakened Claims

The possibility of machine morality is highly contentious. Over the last twenty years, the ethics of artificial intelligence has transformed significantly. Following an increased research focus on artificial general intelligence (or AGI) and speculation of an impending "technological singularity,"[28] there was an increased interest in the possibility of artificial moral agency, moral standing, and concerns around the existential risk machine intelligence poses to humanity. However, with the lack of progress in AGI research and an increasing awareness of the complexity of machine morality, there has been a definite trend away from such anticipatory research topics. Instead, applied ethical issues regarding existing technologies, such as data bias and privacy, have received increasing attention.

In this less favorable research environment, meaningful developments in machine morality are inhibited by theoretical arguments working with narrow definitions of moral concepts that profit from the rich network of associations connected to the concept. This raises a general suspicion of such theoretical claims, leading to an increased skepticism about the possibility of real progress.

An example is Floridi's deflated conception of artificial moral agency. He considers an artificial agent (interactive, autonomous, and adaptable) to be moral if it can act in a way that could "cause moral good or evil"[29] and be measured against a "morality threshold."[30] His definition does not require any form of moral deliberation or choice – capacities typically considered an essential part of any demanding form of moral agency.

Missed Opportunities

The deontic bias results in potentially rewarding avenues for future research being overlooked. The discussion on machine supererogation illustrated this point. By looking at the idea of machine morality from the point of view of supererogation, issues requiring further clarification were identified, such as the lack of an established baseline morality for artificial moral agents. Additionally, several ideas worthy of further investigation were raised, such as applying axiological and aretaic concepts to machine morality and examining the tension between morality's social and personal dimensions in machine agents.

The point is that it is easier to join existing debates than establish new ones within an academic discipline. The present bias toward social morality and deontic concerns of obligation and prohibition has significant momentum, focusing the discussion on certain topics, such as moral agency/patiency and AI alignment. At the same time, other worthwhile research questions concerning the aspirational face of morality remain largely ignored.

Oversimplification

The final consequence is that, as an interdisciplinary field, moral philosophy is done a disservice when moral theory is portrayed in an oversimplified manner. Within machine ethics, it is acceptable to work with narrow conceptual interpretations (when clearly stated) as we can justifiably assume an awareness of alternatives and the complexities of moral theory. However, when presenting such work to members of other disciplines, it is important that claims are sufficiently explained and qualified so that an awareness of the underlying conceptual complexity is communicated. Computer scientists, cognitive scientists, product designers, legal scholars, and public administrators need to understand that moral theory is much more than just the deontic concerns of social morality. Simplifying assumptions and trade-offs may be unavoidable realities, but they should never be interpreted as theoretically comprehensive.

The "Moral Machine Project"[31] is an interesting example of this oversimplification issue. It is to be commended that in pursuing their overall project goal (to put "society in the loop"[32]) the research team clearly acknowledge the limits of their moral dilemma experiments. They actively engage with academic criticism, while emphasizing that their work pursues the realistic objective of identifying generalized principles reflecting the moral intuitions of the general public. Even with this explicit acknowledgment of scope limits and academic criticism, it has still developed into a "poster child"[33] for the ethical issues arising with automated vehicle research. There are similarities to the connection between Asimov's laws of robotics and the AI control problem – both are a double-edged sword. On one hand, they raise awareness, helping non-ethicists to engage with such ethical issues.

On the other, they present a very contrived form of ethical dilemma that ignores significant complexity and nuance in moral theory. The oversimplification issue is comparable.

CONSTRUCTIVE SUGGESTIONS

Finally, three constructive suggestions are made for countering this deontic bias.

Refocus the Debate

One objective of this chapter is to raise awareness of the issue. Only then is it possible to consider how such a bias has shaped the machine ethics debate, leading researchers to focus on certain issues, while ignoring others. Some topics in machine morality deserve less attention – they have a fixed ceiling due to the conceptual limits of deontic moral theory. Others remain unexplored, yet they have the potential to significantly progress our thinking on machine morality. The conceptual analysis of machine supererogation is an example. At the least, researchers should pause to evaluate the limits of their conceptual assumptions and consciously decide to work within them.

Default Skepticism

Irrespective of how optimistic one is about the possibility of machine morality, a default skepticism is counseled when engaging with machine ethics research. It is asymmetrical – machine ethicists typically work with narrowly scoped moral theory lacking the conceptual richness of human morality. The inverse does not presently exist. This is not skepticism about the possibility of making real progress in the field. It simply recommends critically examining the conceptual limits of any underlying moral theory so that the significance of any claims can be accurately qualified.

The Pedagogical Challenge

Such an interdisciplinary field poses a significant pedagogical challenge for machine ethicists. There are fundamental moral issues deserving both resources and attention. Other disciplines need to understand and engage with these problems. The challenge is to meet a discursive "standard of substantial completeness"[34] in which all relevant information is communicated to other parties, thereby facilitating an accurate understanding of the problem, while not overwhelming them with unnecessary complexity or detail.

CONCLUSION

Machine ethics presently has a deontic bias. Conceptually, our academic debate is narrowly focused on the application of duties and rights to meet concerns of social morality. This undermines claims about the possibility of machine morality, while also overlooking potentially rewarding avenues for future research. In addition, it presents moral theory in an oversimplified light, leading to a lack of philosophical depth in interdisciplinary research.

The first half of this paper explored the idea of machine supererogation. This served two purposes. First, it showed the limitations of a purely deontic interpretation of morality. Three specific concerns were raised – the difficulty in establishing a baseline of minimal duties for machine intelligence, the value of *licentia* for a moral agent, and the tension existing between the social and personal dimensions of morality. Second, it demonstrated the potential for new insights to be gained by exploring the connection between machine intelligence and the aspirational side of morality.

The second half explained the negative impact of the deontic bias in machine ethics. Parallels to Sharkey's "linguistic trojan horse" concept were noted in that claims based on narrow conceptual definitions profit from their association with broader, more nuanced interpretations. This leads to the negative consequences outlined above – weaker claims, missed opportunities, and a theoretical shallowness to moral theory within interdisciplinary contexts. With an increased awareness of this issue, machine ethics debate should be refocused to profit from the full conceptual depth of moral theory.

There is good reason to be optimistic about machine ethics. The issues it grapples with are important. Progress in the field will not only shape the direction of artificial intelligence research but also help define the future role of AI within society. This chapter argued that its current deontic bias is detrimental to machine ethics. It distorts debate, resulting in effort being invested disproportionately into the concerns of social morality, while largely ignoring the personal and aspirational components of morality. To make real progress, we need a lucid recognition of the limits deontic bias imposes on our thinking. But, more importantly, we need to be open to exploring alternative approaches to viewing machines as moral subjects.

NOTES

1 Vincent Müller, "Ethics of Artificial Intelligence and Robotics," in *The Stanford Encyclopedia of Philosophy*, ed. Edward Zalta (Stanford: Stanford University, Summer 2020 Edition), 29.

2 Jan Leike, Miljan Martic, Victoria Krakovna, Pedro A. Ortega, Tom Everitt, Andrew Lefrancq, Laurent Orseau, and Shane Legg, "AI Safety Gridworlds," *arXiv:1711.09883* (2017): 2.

3 David Heyd, *Supererogation: Its Status in Ethical Theory* (Cambridge: Cambridge University Press 1982), 10.

4 James Ople Urmson, "Saints and Heroes," in *Moral Concepts,* ed. Joel Feinberg (Oxford: Oxford University Press, 1960), 65.

5 David Heyd, *Supererogation: Its Status in Ethical Theory* (Cambridge: Cambridge University Press 1982), 115.

6 James Ople Urmson, "Hare on Intuitive Moral Thinking," in *Hare and Critics: Essays on Moral Thinking,* ed. D. Seanor and N. Fotion (Oxford: Oxford University Press, 1988), 169.

7 David Heyd, "Supererogation," in *The Stanford Encyclopedia of Philosophy*, ed. Edward Zalta (Stanford: Stanford University, Spring 2016 Edition), 3, https://plato.stanford.edu/archives/spr2016/entries/supererogation/.

8 James Ople Urmson, "Saints and Heroes," in *Moral Concepts,* ed. Joel Feinberg (Oxford: Oxford University Press, 1960), 64.

9 Gregory Mellema, "Moral Expectation," *The Journal of Value Inquiry* 32, no. 4 (1998): 479–488.

10 Gregory Trianosky, "Supererogation, Wrongdoing, and Vice: On the Autonomy of the Ethics of Virtue," *The Journal of Philosophy* 83, no. 1 (1986): 28.

11 Eric Dietrich, "Homo Sapiens 2.0: Building the Better Robots of Our Nature," in *Machine Ethics,* ed. Michael Anderson and Susan Leigh Anderson (New York: Cambridge University Press, 2011), 537.

12 Susan Wolf, "Moral Saints," *The Journal of Philosophy* 79, no. 8 (1982): 419.

13 Drew McDermott, "What Matters to a Machine," in *Machine Ethics,* ed. Michael Anderson and Susan Leigh Anderson (New York: Cambridge University Press, 2011), 108.

14 David Heyd, "Supererogation," in *The Stanford Encyclopedia of Philosophy*, ed. Edward Zalta (Stanford: Stanford University, Spring 2016 Edition), 23.

15 Nick Bostrom, *Superintelligence: Paths, Dangers, Strategies* (Oxford: Oxford University Press, 2014), 127–144.

16 Donald Crosby, *The Specter of the Absurd: Sources and Criticisms of Modern Nihilism* (Albany: State University of New York Press, 1988), 15.

17 David Levy, "Assimilating Supererogation," in *Supererogation – Royal Institute of Philosophy Supplement 77,* ed. Christopher Cowley (Cambridge: Cambridge University Press, 2015), 239.

18 Elizabeth Drummond-Young, "Is Supererogation More Than Just Costly Sacrifice?" in *Supererogation – Royal Institute of Philosophy Supplement 77,* ed. Christopher Cowley (Cambridge: Cambridge University Press, 2015), 137.

19 Millard Schumaker, "Deontic Morality and the Problem of Supererogation," *Philosophical Studies* 23, no. 6 (1972): 428.

20 Noel Sharkey, "The Evitability of Autonomous Robot Warfare," *International Review of the Red Cross* 94, no. 886 (2012): 793.

21 Bertram Malle, "Integrating Robot Ethics and Machine Morality: The Study and Design of Moral Competence in Robots," *Ethics and Information Technology* 18, no. 4 (2016): 246.

22 Joanna Bryson, "Patiency is not a Virtue: The Design of Inelligent Systems and Systems of Ethics," *Ethics and Information Technology* 20, no. 1 (2018): 16.

23 Joanna Bryson, "Patiency is not a Virtue: The Design of Inelligent Systems and Systems of Ethics," *Ethics and Information Technology* 20, no. 1 (2018): 22.

24 Patrick Lin, "Why Ethics Matters for Autonomous Cars," in *Autonomous Driving: Technical, Legal and Social Aspects,* ed. Markus Maurer, J. Christian Gerdes, Barbara Lenz, and Hermann Winner (2016), 69–86.

25 Wulf Loh and Janina Loh, "Autonomy and Responsibility in Hybrid Systems: The Example of Autonomous Cars," in *Robot Ethics 2.0: From Autonomous Cars to Artificial Intelligence,* ed. Patrick Lin, Keith Abney, and Ryan Jenkins (Oxford: Oxford University Press, 2017), 35–50.

26 Noel Sharkey, "The Evitability of Autonomous Robot Warfare," *International Review of the Red Cross* 94, no. 886 (2012): 793.

27 Michael Ferry, "Beyond Obligation: Reasons and Supererogation," in *Supererogation – Royal Institute of Philosophy Supplement 77,* ed. Christopher Cowley (Cambridge: Cambridge University Press, 2015), 64.

28 Ray Kurzweil, *The Singularity is Near: When Humans Transcend Biology* (London: Viking, 2005).

29 Luciano Floridi, *The Ethics of Information* (Oxford: Oxford University Press, 2013), 147.

30 Luciano Floridi, *The Ethics of Information* (Oxford: Oxford University Press, 2013), 152.

31 Edmond Awad, Sohan Dsouza, Richard Kim, Jonathan Schulz, Joseph Henrich, Azim Shariff, Jean-François Bonnefon, and Iyad Rahwan, "The Moral Machine Experiment," *Nature* 563, no. 7729 (2018): 59–64.

32 Edmond Awad, Sohan Dsouza, Jean-François Bonnefon, Azim Shariff, and Iyad Rahwan, "Crowdsourcing Moral Machines," *Communicaions of the ACM* 63, no. 3 (2020): 55.

33 Edmond Awad, Sohan Dsouza, Jean-François Bonnefon, Azim Shariff, and Iyad Rahwan, "Crowdsourcing Moral Machines," *Communicaions of the ACM* 63, no. 3 (2020): 51.

34 Tom Beauchamp and James Childress, *Principles of Biomedical Ethics* (New York: Oxford University Press, 2001), 288.

REFERENCES

Awad, Edmond, Sohan Dsouza, Jean-François Bonnefon, Azim Shariff, and Iyad Rahwan, "Crowdsourcing Moral Machines," *Communicaions of the ACM* 63, no. 3 (2020): 48–55.

Awad, Edmond, Sohan Dsouza, Richard Kim, Jonathan Schulz, Joseph Henrich, Azim Shariff, Jean-François Bonnefon, and Iyad Rahwan, "The Moral Machine Experiment," *Nature* 563, no. 7729 (2018): 59–64.

Beauchamp, Tom, and James Childress, *Principles of Biomedical Ethics* (New York: Oxford University Press, 2001).

Bostrom, Nick, *Superintelligence: Paths, Dangers, Strategies* (Oxford: Oxford University Press, 2014).

Bryson, Joanna, "Patiency is not a Virtue: The Design of Inelligent Systems and Systems of Ethics," *Ethics and Information Technology* 20, no. 1 (2018): 15–26.

Crosby, Donald, *The Specter of the Absurd: Sources and Criticisms of Modern Nihilism* (Albany: State University of New York Press, 1988).

Dietrich, Eric, "Homo Sapiens 2.0: Building the Better Robots of Our Nature," in *Machine Ethics*, ed. Michael Anderson and Susan Leigh Anderson (New York: Cambridge University Press, 2011), 531–538.

Drummond-Young, Elizabeth, "Is Supererogation More Than Just Costly Sacrifice?" in *Supererogation – Royal Institute of Philosophy Supplement 77*, ed. Christopher Cowley (Cambridge: Cambridge University Press, 2015), 125–141.

Ferry, Michael, "Beyond Obligation: Reasons and Supererogation," In *Supererogation – Royal Institute of Philosophy Supplement 77*, ed. Christopher Cowley (Cambridge: Cambridge University Press, 2015), 49–66.

Floridi, Luciano, *The Ethics of Information* (Oxford: Oxford University Press, 2013).

Heyd, David, *Supererogation: Its Status in Ethical Theory* (Cambridge: Cambridge University Press 1982).

Heyd, David, "Supererogation," in *The Stanford Encyclopedia of Philosophy*, ed. Edward Zalta (Stanford: Stanford University, Spring 2016 Edition), https://plato.stanford.edu/archives/spr2016/entries/supererogation/.

Kurzweil, Ray, *The Singularity is Near: When Humans Transcend Biology* (London: Viking, 2005).

Leike, Jan, Miljan Martic, Victoria Krakovna, Pedro A. Ortega, Tom Everitt, Andrew Lefrancq, Laurent Orseau, and Shane Legg, "AI Safety Gridworlds," *arXiv - 1711.09883* (2017): 1–20.

Levy, David, "Assimilating Supererogation" in *Supererogation – Royal Institute of Philosophy Supplement 77*, ed. Christopher Cowley (Cambridge: Cambridge University Press, 2015), 227–243.

Lin, Patrick, "Why Ethics Matters for Autonomous Cars," in *Autonomous Driving: Technical, Legal and Social Aspects*, ed. Markus Maurer, J. Christian Gerdes, Barbara Lenz, and Hermann Winner (Berlin, Heidelberg: Springer, 2016), 69–86.

Loh, Wulf, and Janina Loh, "Autonomy and Responsibility in Hybrid Systems: The Example of Autonomous Cars," in *Robot Ethics 2.0: From Autonomous Cars to Artificial Intelligence*, ed. Patrick Lin, Keith Abney, and Ryan Jenkins (Oxford: Oxford University Press, 2017), 35–50.

Malle, Bertram, "Integrating Robot Ethics and Machine Morality: The Study and Design of Moral Competence in Robots," *Ethics and Information Technology* 18, no. 4 (2016): 243–256.

McDermott, Drew, "What Matters to a Machine," in *Machine Ethics*, ed. Michael Anderson and Susan Leigh Anderson (New York: Cambridge University Press, 2011), 88–114.

Mellema, Gregory, "Moral Expectation," *The Journal of Value Inquiry* 32, no. 4 (1998): 479–488.

Müller, Vincent, "Ethics of Artificial Intelligence and Robotics," in *The Stanford Encyclopedia of Philosophy*, ed. Edward Zalta (Stanford: Stanford University, Summer 2020 Edition), 1–70.

Schumaker, Millard, "Deontic Morality and the Problem of Supererogation," *Philosophical Studies* 23, no. 6 (1972): 427–428.

Sharkey, Noel, "The Evitability of Autonomous Robot Warfare," *International Review of the Red Cross* 94, no. 886 (2012): 787–799.

Trianosky, Gregory, "Supererogation, Wrongdoing, and Vice: On the Autonomy of the Ethics of Virtue," *The Journal of Philosophy* 83, no. 1 (1986): 26–40.

Urmson, James Ople, "Saints and Heroes," in *Moral Concepts,* ed. Joel Feinberg (Oxford: Oxford University Press, 1960), 60–73.

Urmson, James Ople, "Hare on Intuitive Moral Thinking," in *Hare and Critics: Essays on Moral Thinking,* ed. D. Seanor and N. Fotion (Oxford: Oxford University Press, 1988), 161–169.

Wolf, Susan, "Moral Saints," *The Journal of Philosophy* 79, no. 8 (1982): 419–439.

CHAPTER 6

The Hacker Way

Moral Decision Logics with Lethal Autonomous Weapons Systems

Elke Schwarz

INTRODUCTION

The last decade has seen a heightened engagement with a technology that is already more than half a century old. Artificial Intelligence, as a field of scientific research and as a concept, has its origins in the 1950s with Alan Turing's landmark text, "Computing Machinery and Intelligence"; and with ideas about automatic, self-regulating control taking shape within the field known as cybernetics.[1] In 1956, the term "Artificial Intelligence" (AI) was coined, as part of a two-month long workshop in Dartmouth, organized by John McCarthy and attended by fellow pioneers Marvin Minsky, Nathan Rochester and Claude Shannon. Since then, the field has seen various waves of innovation in which new technological capacities advanced the possibility of using sophisticated AI to produce a "thinking machine." In the early days, the idea that AI would make a substantial contribution to the practice of warfare in ways that might prove challenging was already articulated by some of the pioneers of cybernetics. Perhaps the most vocal among them was Norbert Wiener, who was under no illusion that the abstract field of mathematics, on which forays in cybernetics and AI rest, might well be "as dangerous as a potential armourer of the new scientific war of the future."[2]

Despite being an eminent scientist of technology and mathematical thinker, Wiener was ill at ease with the increasingly prevalent perspective which viewed technology "not so much as applied science, but rather as applied social and moral philosophy."[3] In other words, Wiener saw what would become less visible as the digital technological substrate became increasingly dense and ubiquitous: that the logic of computation and information

DOI: 10.1201/9781003226406-7

technology has become a dominant mode of philosophical reasoning and is set to shape ethical thinking in significant ways. This confuses considerations about what is good for society with considerations about what tools and technologies we can make for society, and prioritizes our fascination with how these technologies work over ethical considerations more generally. In Wiener's words, a key challenge of modernity is the "worship of *know-how*, as opposed to *know-what*."[4] That is to say that we tend to be inclined to "accept the superior dexterity of the machine-decision without too much inquiry as to the motives and principles behind these."[5] For Wiener, this was a problematic oversight which comes at the expense of our collective social benefit – the know-what "by which we determine not only how to accomplish our purposes, but what our purposes are to be."[6] This highly normative question – what our purposes are to be – risks becoming subsumed into technical jargon and processes. This condition is palpable in the debates on Lethal Autonomous Weapons Systems (LAWS) today, where technological development outpaces policy deliberations quite significantly.

In this chapter, I follow Norbert Wiener and argue that, in the fervent pursuit of military AI for increasingly autonomous weapons systems, we have indeed lost sight of the *know-what* in favor of a blinkered focus on *know-how*. This puts in jeopardy not only our global security, but the very integrity of our human ethical reasoning and the ability, and willingness, to take responsibility for our technologically mediated acts of violence – intended or accidental. Indeed, we have much to learn from the early thinkers of autonomy in weapons systems and their emphatic warning that our humanity is closely tied to our ability to have and take responsibility. The chapter begins by tracing the rapid forays into full autonomy in weapons systems, including human targeting, for which AI plays an increasingly crucial role, before exploring how the contemporary entwinement of the private technology industry with military operations problematically prioritizes know-how over know-what. The final section then focuses on the key tension between technological functioning and social consciousness, arguing that the prevalence of machine logics implicit in the know-how perspective is likely to lessen the restraint on harm in warfare for the foreseeable future.

THE RAPID RISE OF LAWS

Today, AI is cast as a panacea for a wide range of fields, from agriculture and business to healthcare, the justice system, and our day-to-day environment. Indeed, narrow AI is already employed within many of these fields today. It is not surprising then that AI is set to play an increasingly significant role for military operations as well. Those states or organizations that have published specific military AI strategies typically stress two aspects from the outset: first, that AI is set to change the way wars are fought in a significant manner, and second, that AI will yield tremendous benefits to military organizations across a range of domains, such that those who lead in the field will acquire a significant competitive advantage. Whether and how AI will indeed alter the nature of warfare is at this stage is still speculative. That AI may bring considerable advantages in terms of speed, efficiency and process optimization is clear. Less obvious is whether amplified investment in AI might yield outcomes that exacerbate or mitigate the potential for more violence in warfare. The promise of a faster and cognitively superior weapons system has a strong allure. The question, however, is: to what end? The desire for greater automation and the trajectory toward autonomy in machine-decision-making for military purposes has, like AI itself, a longer history. In his seminal text *Die Antiquiertheit des Menschen* (The Obsolescence of Man), published in 1956, Günther Anders tells the story of General Douglas MacArthur, who was discharged from his duties as United Nations commander in the Korean War, on account of his rather hawkish approach. MacArthur's strategic decisions capacities, Anders recalls, were superseded not by a different General or a team of individuals, but rather by the latest military technology, a so-called "Electric Brain"; a technological object that would be able to determine with greater objectivity what the best way forward might be.[7] While Anders indicates that this was a well-known story at the time, the details of the exact make or model of the machine are difficult to trace. It is possible that Anders referred to IBM's 701 model, also known as the "Defence Calculator," which was conceived in response to problems arising out of the Korean War and was first used by the US Navy in the early 1950s;[8] or perhaps Anders' "Electric Brain" was a reference to the ERA 1101/Atlas model, the first stored-programmed computer, and also one of the early "thinking machines" designed initially for the US Navy.[9] The point of the story in Anders' account was to raise alarm bells that lethal decision-making should never be delegated to a machine. For Anders, the substitution of human decision making with machine-decisions in matters of war constitutes a grave moral challenge, wherein we sacrifice both our conscience and our agency "at the altar of the apparatus," ready to submit to the machine and accept it as a proxy conscience and oracle, as a "prediction machine" to mitigate our own moral risk.[10]

What might have been an exceptional and rudimentary military venture in Anders' time is fast becoming a normality in our contemporary context, as a growing number of systems emerge that replace human decision making with the superior cognitive capacities of AI-enabled autonomous systems in all manner of tasks including the autonomous identification, tracking and attacking of enemy targets, including human targets. Of concern to most critics of lethal autonomy in weapons systems are not the oft-invoked dystopian Terminator-style Killer Robots, but rather the more mundane yet perhaps more dangerous systemic assemblages by which an AI decision system is paired with a sensor-equipped weapons platform – such as a drone – to identify and attack targets autonomously throughout the entire kill-chain. The United States' forays into lethal autonomy are illustrative of the general push for more military AI use in targeting decisions. In 2017, it pioneered its Algorithmic Warfare Cross-Function Team, better known as Project Maven. Project Maven's main aim is to harness the benefits of machine learning and computer vision to accelerate targeting (and other) decisions in challenging conflict environments. The project draws on drone technology for its visual and other sensory inputs to create the ability to identify and track potential targets in real-time to give the human intelligence for kinetic engagement at speed. Since then, various systems of this nature have been trialed, tested and designed to meet the US military's objective of achieving "decision dominance" with faster and more lethal action through AI.[11] In August of 2020, DARPA flew a trial mission in which "several dozen military drones and tank-like robots took to the skies and roads"[12] with the aim of tracking down terrorist suspects in an urban environment. This was just one of many such exercises that indicates the Pentagon's determination to shift an increasing amount of decision-making, including lethal decision-making, to machines. Indeed, the suggestion put forward by General John Murray, among others, is that we may perhaps not need any human control of lethal AI systems after all, since that would impeded the superior capacities of the technologies available.[13] The assumption is that the system can do it faster and more comprehensively and that the human presents an obstacle in this process. This represents a considerable shift in the narrative put forward by the US Department of Defense which, until now, had strongly emphasized that adequate human control in critical targeting functions would always remain a requirement in the operation of autonomous weapons systems.[14] It is clear then that the goal posts are now shifting considerably and quite quickly. As was the case with drone use in counter-terrorism operations, the US sets the tone for the modes of warfare to come, and AI-enabled anti-personnel targeting weapons that will operate without human oversight are clearly on the near horizon.

It was, however, a Turkish autonomous drone system – the STM KARGU-2 – that made significant ripples in 2021, when it was mentioned in a UN Report by a panel of experts on Libya as having been used in 2020 to "hunt down and remotely engage" one of the parties to the hostilities. The report clearly names the system as a LAWS as it notes that "[t]he lethal autonomous weapons systems were programmed to attack targets without requiring data connectivity between the operator and the munition."[15] The full details on the incident are scant – it is not clear whether there was some level of human oversight, whether it was a regular type of loitering munition, or whether anyone came to harm through the system. However, if it was indeed used in full autonomy mode to identify, track and attack human targets, this would – at the time of writing – be the first publicly documented anti-personnel use of a LAWS. The STM KARGU is a small quad-copter, built by the Turkish defense company STM, which "employs machine-learning algorithms embedded on the platform" and is designed specifically for "asymmetric warfare or anti-terrorist operations."[16] The manufacturer's claims as to the sophistication of the technology are somewhat vague and likely to be exaggerated. Many would argue that the state of the art of technological development is nowhere near being able to distinguish between friend or foe in the messy reality of the battlefield.[17] Systems that employ AI for the full kill-chain are likely to be marred by incomplete, low-quality, incorrect or discrepant data.[18] This, in turn, will lead to highly brittle systems and biased, harmful outcomes. As UNIDIR researcher Michael Holland notes: "conflict environments are harsh, dynamic and adversarial, and there will always be more variability in the real-world data of the battlefield than in the limited sample of data on which autonomous systems are built and verified."[19] Moreover, data for AI is often mislabelled,[20] and research in machine learning for AI is by far not as scientifically robust as it might appear from media accounts.[21] This condition will inevitably lead to faulty decisions, to misapplications of force and other harmful errors. In short, accidents will happen with the use of AI-enabled lethal autonomous systems for anti-personnel targeting. This much we know. I will return to the category "accident" as an important issue of moral significance in the final section below.

Where accidents are likely to happen, it is necessary that someone ought to be able to take or be assigned responsibility, so that accidents do not become a regular harmful practice. The question of responsibility is of course notoriously thorny, and much has been written to explain the responsibility gap that emerges with LAWS over the last fifteen years.[22] What complicates matters in the assignment of responsibility and the regulation of AI-enabled autonomous systems is the very *systemic* nature of such weapons, the dual-use character of AI components and the fact that AI technologies and systems are no longer developed and provided with relatively close governmental oversight, or indeed through governmental agencies, but are produced by and acquired primarily from private technology companies. In other words, there is now a highly complex and firmly entwined relationship between Silicon Valley interests and military progress. Former Google CEO Eric Schmidt is driving the military AI agenda forward with scarcely mitigated enthusiasm and his influence can be felt in the relatively recent shift to prioritize speed in action and innovation as a core principle for the future of the US military. The dynamics of this relationship and the increasingly notable dominance of private sector rhetoric for military operations, should, however, give us pause for thought. What happens when the military picks up the pace of the technology industry? Are these two entities and two very different purposes compatible? With an enormous push for more AI in military operations, the prevailing aim seems to be the development of ever more intricate technological know-how, much to the detriment of thinking more clearly about the purposes and aims of what we want for society, or indeed for the communities that are subject to war.

SILICON VALLEY'S MILITARY ADVANCE

The market for military AI is lucrative. In 2020, the overall AI in military market was worth US$ 6.3 billion and it is forecast to nearly double in the next five years to US$11.6 billion. This rise is led largely by the US, which has been increasing expenditure on AI year on year to push forward its competitive edge.[23] Among the companies keen to establish their military AI products are key global players in the defense industry like BAE Systems PLC, Northrup Grumman Companies, and Lockheed Martin Corporation. But other, non-defense specific actors are taking on an increasingly important role in supplying computer vision, machine learning and other AI system elements. Moreover, the technology industry has been quite forcefully promoting the increased use of AI in the military in the past five years. Perhaps the most well-known among these technology companies which have crossed over from consumer information technology provider to military contractor is Google. The multi-billion technology company made headlines in 2018 when Google employees publicly registered their dissent in providing AI technologies that could potentially be used for autonomous targeting for the Pentagon's Project Maven.[24] The ensuing public outcry was sufficient to compel Google to not renew its contract for the Project. However, smaller technology companies with fewer qualms about public sentiments were quick to step in. One of these companies is Anduril Industries, a new company started in 2017 by the founder of Oculus Rift, a company pioneering VR technology for gaming. Anduril Industries has fashioned itself as a defense technology company, although its ties with defense are still young. Nonetheless, the company's ambitions are clearly articulated by a Tweet from its founder, Palmer Luckey, "to turn American and allied warfighters into invincible technomancers who wield the power of autonomous systems to safely accomplish their mission."[25] For the technology industry, the defense sector is the next lucrative frontier and the ethical dimensions of producing technologies that are potentially used for lethal operations are less significant for many Silicon Valley engineers. As Luckey quips, "most engineers want to engineer. They want to get stuff done."[26]

This enmeshment between new technology start-ups and the military raises obvious security issues at the global level that are in urgent need of regulatory address. One of these is the multi-national character of such companies. Anduril, for example, prides itself in "actively supporting operations with the [U.S.] Department of Defense, the U.S. Department of Homeland Security, and the UK Ministry of Defence, as well as other agencies" with their Lattice AI products.[27] Similarly, another technology company with an equally ambitious international expansion agenda – Palantir – is making headway in providing their AI products to multiple states' defense agencies, including the UK's MoD and the US DoD. The technology sector is not nearly as well-regulated as the defense industry, and there is understandable concern about national data privacy, export regulations for weapons system components, safety standards of products, a lack of standards and protocols for use of military AI, a shortage of trained personnel, and so on. Technology companies, although tasked with handling sensitive military data, are not beholden to national security; they are primarily beholden to their stakeholders. The area is highly opaque and entirely lacking in oversight.

To be clear, technological innovations and military endeavors have long gone hand in hand. Wiener's innovations in communication theory were used for missile technologies in his time, the pioneers of AI almost all worked on mid-century military projects, and of course the Internet, now a ubiquitous technology across the globe, emerged entirely from a military endeavor. Traditionally, it was the requirements of military organizations that would set the pace and structure for technological innovations, and defense companies had, previously, worked in tandem with the military to foster innovation and ensure competitive advantage. The cutting-edge of technological innovation materialized primarily through specialist defense affiliated organizations like the Pentagon's Defense Advance Research Projects Agency (DARPA), tasked with blue sky thinking in military innovation. DARPA is still a major driver of innovation for US defense, however, the as-of-recently commercially oriented Silicon Valley ethos appears to be increasingly in the driving seat when it comes to dictating the rhythm and pace of military technological innovation, the potential use of new products, and the cultural shift toward autonomy in selecting, tracking and taking out targets without a human in the loop. In particular, the Silicon Valley ideology of speed and agility in churning out products is becoming a dominant theme. Facebook, for example, has built its business around the ethos that there is intrinsic value in getting products to market at speed, regardless of whether these technological ideas are fully fit for use. As Mark Zuckerman notes in a 2009 interview, his engineers are encouraged to embrace errors and mistakes – to "break things." Only by embracing the logic of error and mistakes can coders move fast enough to gain an advantage over competitors in the market. As he notes in the interview: "At the end of the day the goal of building something is to build something not to not make mistakes."[28] Shortcuts and *post hoc* fixes are always justified with the understanding that mistakes can always be addressed later – "Done is better than perfect," to quote another Facebook slogan. The business practice is otherwise also known as "The Hacker Way." This mode of thinking is fast becoming an aspiration across industrialized nations and a mode of social, economic, political and indeed military organization.

This is perhaps not surprising. Former Google CEO Eric Schmidt has been a very vocal advocate of making military processes and procedures much faster and he has forged ties with Washington to promote his "Silicon Valley worldview where advances in software and AI are the keys to figuring out almost any issue."[29] For Schmidt, traditional processes and procurement procedures in military operations are far too slow and he appears determined to align the timeframes of action between the military world and the technology world in his push to modernize the US military. The goal, he notes, "should be to have as many software companies to supply software of many, many different kinds: military H.R. systems, email systems, things which involve military intelligence, weapons systems and what have you."[30] Schmidt's efforts seem to be bearing fruit. The technology worldview is finding its way into military and policy circles. US Army Chief of Staff General James McConville declared in early 2021, that the Army must become faster, both "as an institution developing new weapons – not in decades, but in a few years – and as a battlefield force, destroying enemies – firing artillery, not minutes after spotting an enemy, but seconds."[31] In March 2021, Chris Dougherty, Senior Fellow at the US Center for New American Security (CNAS), urged the Pentagon to take a leaf out of the software industry playbook if the Pentagon wants to become truly agile. He recommends the adoption of a concept from the software industry – agile development – for both acquisition and strategy. Agile software development is built on an number of principles that foreground dynamic and iterative processes. This means that products get developed and delivered quickly and then refined, improved, adjusted and fine-tuned as time goes on. This, Dougherty reasons, should be the strategic way forward for the Pentagon: "why not issue something quickly – call them "minimum viable strategies" – comprising critical priorities and tradeoffs early in an administration, then iterate on these strategies over time?"[32]

Whether this is indeed an ethically viable concept for the strategic development and implementation of lethal targeting technologies stands to reason. It is clear, then, that the culture of military acquisition, as well as formulating strategy is subject to influence from the technology sector's advances. The ideology and pace of this industry is, however, also increasingly shaping ideas of critical decision-making in military tactics – including lethal targeting decisions. In other words, the Hacker Way, where decisions grounded in technological functionality are taken quickly with the assumption that errors and problems can then be fixed later, is making forays into matters of life and death. The technology industry's influence in changing the culture on autonomous killing with AI systems is, perhaps, best illustrated by another project chaired by Eric Schmidt – the National Security Commission on AI (NSCAI) Report, issued on March 1, 2021. The report presents the Commissions "strategy for winning the AI era"[33] and is the result of a prolonged period of discussion and consultation with various stakeholders, including the military. The first part of the report is dedicated to "Defending America in the AI Era" and includes strategic suggestions for the development for lethal autonomous weapons systems. The recommendation of the Commission is clear: "[t]he Department [of Defense] must act now to integrate AI into critical functions, existing systems, exercises and wargames to become an AI-ready force by 2025."[34] That AI compresses decision times from minutes to seconds, and that this is likely to exceed any human decision capabilities is a given.[35] The fact that the US government "still operates at human speed, not machine speed" is seen as a considerable detriment to progress and a disadvantage that must be overcome.[36] To solve this problem, more technology is offered. The AI-promise is that "a machine can perceive, decide and act more quickly, in a more complex environment, with more accuracy than a human."[37] Leaving aside for a moment that experts agree that no AI can operate reliably in a complex conflict environment in this manner, it is clear to see how human reasoning and decision-making is side-lined in this report. This shift is notable in the US rhetoric on the role of the human in or on the loop of LAWS. It has, to date, been the US's official position that a human must always be exercising appropriate control over the use of force in the context of autonomous weapons systems.[38] However, since the release of the NSCAI's final report, the tone is changing. Here I come back to General John Murray, who is falling in line with the AI-promise for military targeting and asks: "is it even necessary to have a human in the loop?"[39] The goalposts are shifting toward an acceptance of fully autonomous lethal machine decisions.

Where the tech sector dictates the pace and culture of innovation, military organizations must adjust. Silicon Valley start-up culture operates at a much faster pace than the structurally complex, highly bureaucratic and decidedly hierarchical lines of decision-making typical of military organizations, and with much less concern for ethical implications that arise in the use of new and emerging technologies.[40] At the forefront in pushing the military AI agenda is the desire to facilitate unimpeded technological innovation (often disguised by a rhetoric of the need to keep up with China). Prioritized as a core value is the development of ever more intricate know-how as a clear stand in for know-what. The promise of AI as a panacea for better lethal action rests on the speculative idea that with more data and better know-how, we will somehow solve the problem of war. Yet, as highlighted above, to most observers, it is clear that unforeseen errors – or accidents – are built into any AI-enabled system, which is tasked with operating within a highly dynamic and complex combat environment. This is why civil society organization and critics of LAWS call for an urgent halt to the development of such weapons and for appropriate regulation of the technology. However, there is a clear and mounting tension between the pace and ideology of technological innovation and our human institutions which provide the foundations for social, political and military practices. By nature and design, the law as an institution is, so much slower than business cycles and the demands for innovation in the technology industry. When the military adopts this pace of change and innovation for its practices, ethical and legal oversight over lethal practices in military operations necessarily suffer. As Will Roper, former, Assistant Secretary of the US Air Force for Acquisition, Technology and Logistics, notes: "Once we get [an airborne AI programme] I imagine the technology will move so quickly our policies will have a hard time catching up to it."[41] This seems to be accepted as collateral damage, the price we have to pay for unimpeded progress, and a competitive edge over global competitors such as China. Hyperbole and narratives about the potency – military and otherwise – of AI aside, the growing ubiquity of AI applications across military, social and political fields of use organizes our present and future habitus in ways that affect military ethical practices. Not only through the implicit ethos of speed as a core value, but through a specific logic upon which AI output rests.

THE HACKER WAY: ACCIDENTS WILL HAPPEN

To better grasp the broader impact of the digital military infrastructures we are building, we must then take the logic of the technologies and the industries that produce these technologies seriously. There is always more to a technology than its functioning as a mere tool. This insight is not new. Critical theorists of technology in the 1950s and 1960s located technological rationality at the core of modern social and political life.[42] In such accounts, technology is understood as a logic of organization and control; "not as a means, but ... an environment and a way of life."[43] In this light, technological artifacts cannot be considered neutral, because they always embody particular interests, ideas, values, and dominant perspectives which circulate as superseding reason within society. Langdon Winner, for example, points to a dominant ideology of technology as progress, a modern dogma which occludes a sober look not only at the ways in which social policy and technical means are entwined, but also obstructs a discussion about the "explicit and covert ideas that inevitably reside in objects, ideas about membership, gender, social class, access, control, order, community, freedom and many more."[44] The same applies for the military context and we ought to understand the principles of these technologies and their makers if we want to understand where warfare is headed.

Speed, error, and boundless, uninterrupted output is the cultural cornerstone of the wider technology ethos across Silicon Valley.[45] The products that emerge out of the software industry, including AI, rest on the fundamental principles of speed and error. As Matteo Pasquinelli points out, the very foundations of AI as a paradigm are marked by "limits, approximations, biases, errors, fallacies and vulnerabilities."[46] As the technologies become more pervasive, so does the logic of its inner workings. As AI and digital technologies are dissolved "into the bloodstream of society, to the point where digital life becomes second nature,"[47] the structural and logical signature of the technology becomes a substrate, an infrastructure, which mediates and directs actions along its functional lines. To quote Jean-Francois Lyotard, "a technical 'move' is 'good' when it is better and expends less energy than another."[48] Within the logic of AI computation, normative judgments about desirable actions and practices are rendered through the lens of technological functioning. It is not the human that defines horizons of possibility, it is the computer.

When techno-logics supersede policy, harmful outcomes become folded into the technological practice as accidents – the excess of innovation. These accidents, in turn, stand outside of the realm of ethical deliberations, as an aberration, an exception, even though the error, or mistake, with AI technologies is in fact the norm. It is the very signature of the technology to be iterative, and always in need of fixing and improving. Patricia Owens has discussed this status of the "accident" in warfare in broader terms, and it is worthwhile quoting her here at length

The meaning of an accident is never given. A decision to assign the label to an event, with its usually related idea of "no fault," can be contested by different and unequal parties via arguments supporting particular social and ideological ends. "Accidents" may offer support for a variety of interpretations about "what happened," many of which can be difficult to disprove.[49]

This condition is exacerbated with lethal AI technologies and their black box decision-making at accelerated speeds. That this raises moral questions of utmost urgency for the use of AI-enabled lethal autonomous systems is clear. This is precisely the condition of abdicating responsibility for decision-making Günther Anders had been so vocal about in his warnings in 1956.

I have argued elsewhere that agency is severely limited in accelerated human-machine teaming.[50] But here I would like to highlight in particular the morally challenging convenience of these technological affordances when it comes to ethical questions. In the AI-enabled LAWS discussion highlighted above, the machine is cast as a superior decision-making agent, over the slow and deliberative processes implicit in human thought; the AI-promise is that of a superior decision-making agent which inevitably must replace the human in a growing realm of tasks and decisions. As is the case with most of our new digital dependencies, this is born not only out of a sense of inevitable inferiority, but also a notable sense of convenience. The necessarily slower process of ethical deliberation over life and death decisions are no match for the speed of change dictated by the technology sector and technologies themselves. Human deliberation, like ethics, takes time. It is context dependent and relational and asks us to agree to make choices that are in the collective interest, even though this might mean that no optimal solution can be achieved. But this slow deliberate process is vital to maintaining a sense of humanity, even, or rather particularly, in warfare.

CONCLUSION

The hyperbole surrounding the demands for unimpeded development of AI as a panacea for problems in virtually all domains of life is emblematic of a myopic focus on the 'know-how' over the "know-what." An AI that can imitate human dance moves; an AI that promises to identify human emotions; an AI that can compose music; or indeed an AI that … these might constitute technological feats, but to what end? What is the purpose of such systems for society? What values do they support? In explaining America's peculiar and dominant obsession with know-how, Wiener recounts the story of a prominent US engineer who bought a mechanized piano. The purchase was made not out of a love for piano music, but rather because of an overwhelming curiosity about the piano's mechanical functioning. Wiener notes that,

> [f]or this gentleman, the player-piano was not a means of producing music, but a means of giving some inventor the chance of showing how skillful he was at overcoming certain difficulties in the production of music. This is an estimable attitude in a second year high-school student. How estimable it is in one of those on whom the whole culture future of the country depend, I leave to the reader.[51]

A poignant question for our times. When we prioritize technological innovation for innovation's sake, we accept that the logic of technology is setting our values and standards. With this, the human becomes expendable, especially in the context of warfare. An expanded category of "collateral damage" comprises the acceptance of accidents as the necessary excess of technological development. In warfare, unlike in a mundane domestic setting, these accidents, mistakes and errors are likely to cause irreversible harms. The Silicon Valley ethos "Move Fast and Break Things" is a tragic motto for military operations. It side-lines ethics entirely as an element in contemporary warfare. It matters that for such horrendous human activities as warfare, decisions over life and death should be difficult and adhere to slow and stringent demands for deliberation. Taking a life should never be an easy, or indeed systematic decision. Decisions of lethal harms require that we maintain an awareness of *what* we are doing, not merely *how* we are doing it. This is no longer warranted with LAWS for which the human is gradually removed as an agent all together. Shortcuts are the norm in the technological worldview, but they should not become the cornerstone for lethal decisions. The very foundations of any remnant of humanity in warfare rests on this.

NOTES

1 John Brockman, "Introduction," in *Possible Minds: 25 Ways of Looking at AI*, ed. John Brockman (London: Penguin, 2019), 2.
2 Wiener, cited in Steve J. Heims, "Introduction", in *The Human Use of Human Beings: Cybernetics and Society*, ed. Nobert Wiener (London: Free Association Books, 1989), xi.
3 Steve J. Heims, "Introduction", in *The Human Use of Human Beings: Cybernetics and Society*, ed. Nobert Wiener (London: Free Association Books, 1989), xii.
4 Nobert Wiener, *The Human Use of Human Beings: Cybernetics and Society* (London: Free Association Books, 1989), xxx (emphasis added).
5 Nobert Wiener, *The Human Use of Human Beings: Cybernetics and Society* (London: Free Association Books, 1989), 185.
6 Nobert Wiener, *The Human Use of Human Beings: Cybernetics and Society* (London: Free Association Books, 1989), 183.
7 Günther Anders, *Die Antiquiertheit des Menschen: Über die Seele im Zeitalter der zweiten industriellen* Revolution (München: C.H. Beck Verlag, 2010), 60–61.
8 David G. Garson, *Public Information Technology and E-governance: Managing the Virtual State* (Sudbury: Jones and Bartlett Publishers, 2006), 32.
9 David L. Boslaugh, *When Computers Went to Sea: The Digitization of the United States Navy* (Los Alamitos: IEEE Computer Society Press, 1999), 96.
10 Günther Anders, *Die Antiquiertheit des Menschen: Über die Seele im Zeitalter der zweiten industriellen* Revolution (München: C.H. Beck Verlag, 2010), 60; Günther Anders, "Reflections on the H Bomb," *Dissent* 3, no. 2 (1956): 149.
11 Sydney Freedberg Jr., "Army's New Aim is Decision Dominance," *Breaking Defense*, March 17, 2021, https://breakingdefense.com/2021/03/armys-new-aim-is-decision-dominance/.
12 Will Knight, "The Pentagon Inches Toward Letting AI Control Weapons," *WIRED*, May 10, 2021, https://www.wired.com/story/pentagon-inches-toward-letting-ai-control-weapons/.
13 Will Knight, "The Pentagon Inches Toward Letting AI Control Weapons," *WIRED*, May 10, 2021, https://www.wired.com/story/pentagon-inches-toward-letting-ai-control-weapons/.

14 Congressional Research Service, "Defense Primer: U.S. Policy on Lethal Autonomous Weapons Systems," 2020.

15 UN Security Council, "Letter Dated 8 March 2021 from the Panel of Experts on Libya Established Pursuant to Resolution 1973(2011) Addressed to the President of the Security Council," March8, 2021.

16 STM, "KARGU Rotary Wing Attack Drone Loitering Munition System," n.d., https://www.stm.com.tr/en/kargu-autonomous-tactical-multi-rotor-attack-uav.

17 See for example Noel Sharkey, "Guidelines for the Human Control of Weapons Systems," *ICRAC Briefing Paper*, April 2018.

18 Arthur Holland Michel, "Known Unknowns: Data Issues and Military Autonomous Systems," in *UNIDIR*, Geneva: United Nations Institute for Disarmament Research 2021.

19 Arthur Holland Michel, "Known Unknowns: Data Issues and Military Autonomous Systems," in *UNIDIR*, Geneva: United Nations Institute for Disarmament Research 2021, 1.

20 Will Knight, "The Foundations of AI are Riddled With Errors," *WIRED*, March 31, 2021, https://www.wired.com/story/foundations-ai-riddled-errors/.

21 Casey Ross, "Machine Learning is Booming in Medicine. It is Also Facing a Credibility Crisis," *Stat+*, June 2, 2021, https://www.statnews.com/2021/06/02/machine-learning-ai-methodology-research-flaws/.

22 For example, see Robert Sparrow, "Killer Robots," *Journal of Applied Philosophy* 24, no. 1 (2007): 62–77; Andreas Matthias, "The Responsibility Gap: Ascribing Responsibility for the Actions of Learning Automata," *Ethics and Information Technology* 6, no. 3 (2004): 175–183; John P. Sullins, "When is a Robot a Moral Agent?" *International Review of Information Ethics* 6, no. 12 (2006): 23–30; Bhuta, Nehal, Susanne Beck, Robin Geiss, Hin-Yan Liu and Claus Kress (eds.), *Autonomous Weapons Systems: Law, Ethics, Policy* (Cambridge: Cambridge University Press, 2016).

23 ResearchAndMarkets.com, "Global Artificial Intelligence Military Market (2020–2025)," *BusinessWire*, March 23, 2021, https://www.businesswire.com/news/home/20210323005739/en/Global-Artificial-Intelligence-in-Military-Market-2020-to-2025---Incorporation-of-Quantum-Computing-in-AI-Presents-Opportunities---ResearchAndMarkets.com.

24 Scott Shane and Daisuke Wakabayashi, "'The Business of War': Google Employees Protest Work for the Pentagon," *The New York Times*, April 4, 2018, https://www.nytimes.com/2018/04/04/technology/google-letter-ceo-pentagon-project.html.

25 Palmer Luckey, "We Just Raised $450M in Series D funding for Anduril," *Twitter*, June 18, 2021, https://twitter.com/PalmerLuckey/status/1405595273263423490.

26 Luckey, quoted in Cade Metz, "Away from Silicon Valley, the Military is the Ideal Customer," *The New York Times*, February 26, 2021. https://www.nytimes.com/2021/02/26/technology/anduril-military-palmer-luckey.html.

27 Anduril Industries, "Anduril Raises $450 Million in Series D Funding," *Medium*, June 17, 2021, https://medium.com/anduril-blog/anduril-raises-450-million-in-series-d-funding-671f0a27876b.

28 Zuckerberg quoted in Henry Blodget, "Mark Zuckerberg, Moving Fast and Breaking Things," *Business Insider*, October 14, 2010, https://www.businessinsider.com/mark-zuckerberg-2010-10?r=US&IR=T.

29 Kate Conger and Cade Metz, "'I Could Solve Most of Your Problems': Eric Schmidt's Pentagon Offensive," *The New York Times*, May 2, 2020.

30 Schmidt, quoted in Kate Conger and Cade Metz, "'I Could Solve Most of Your Problems': Eric Schmidt's Pentagon Offensive," *The New York Times*, May 2, 2020.

31 Sydney Freedberg Jr., "Army's New Aim is Decision Dominance," *Breaking Defense*, March 17, 2021, https://breakingdefense.com/2021/03/armys-new-aim-is-decision-dominance/.

32 Chris Doughterty, "Want an Agile Pentagon? Don't Go Chasing Waterfalls", *DefenseOne*, March 22, 2021, https://www.defenseone.com/ideas/2021/03/want-agile-pentagon-dont-go-chasing-waterfalls/172835/.

33 National Security Commission on Artificial Intelligence (NSCAI), "Final Report," 2021, https://www.nscai.gov/.

34 National Security Commission on Artificial Intelligence (NSCAI), "Final Report," 2021, 77, https://www.nscai.gov/.

35 National Security Commission on Artificial Intelligence (NSCAI), "Final Report," 2021, 23, https://www.nscai.gov/.

36 National Security Commission on Artificial Intelligence (NSCAI), "Final Report," 2021, 24, https://www.nscai.gov/.

37 National Security Commission on Artificial Intelligence (NSCAI), "Final Report," 2021, 22, https://www.nscai.gov/.

38 Department of Defense, "Department of Defense Directive 3000.09," November 21, 2012, (Incorporating Change 1, May 8 2017), https://www.esd.whs.mil/portals/54/documents/dd/issuances/dodd/300009p.pdf.

39 Murray, quoted in Will Knight, "The Foundations of AI are Riddled With Errors," *WIRED*, March 31, 2021, https://www.wired.com/story/foundations-ai-riddled-errors/.

40 Patrick McGee, "Silicon Valley Reboots its Relationship with the US Military," *Financial Times*, May 17, 2021, https://www.ft.com/content/541f0a02-ea27-43a4-b554-96048c40040d?shareType=nongift.

41 Roper quoted in Harry Lye, "Will Roper on AI and autonomous in the US Air Force," *Airforce Technology*, December 3, 2019, https://www.airforce-technology.com/features/dr-will-roper-on-ai-and-autonomy-in-the-us-air-force/.

42 See for example Herbert Marcuse, "Some Social Implications of Modern Technology," *Studies in Philosophy and Social Science* 9, no. 3 (1941): 414–439; Martin Heidegger, *The Question Concerning Technology and Other Essays* (New York: Harper & Row Publishers, 1977); Günther Anders, *Die Antiquiertheit des Menschen: Über die Seele im Zeitalter der zweiten industriellen* Revolution (München: C.H. Beck Verlag, 2010), 60–61; Jacques Ellul, *The Technological Society* (Toronto: Random House, 1964) for prominent accounts which consider the socio-political power of modern technology.

43 Andrew Feenberg, *Transforming Technology: A Critical Theory Revisited* (New York: Oxford University Press, 1991), 7–8.

44 Langdon Winner, "Three Paradoxes of the Information Age," in *Culture on the Brink: Ideologies of Technology*, ed. Gretchen Bender and Timothy Druckrey (Seattle: Bay Press, 1994), 196.

45 For an interesting survey of how thoroughly this ethos informs Facebook's corporate culture, see Taina Bucher, *IF...THEN: Algorithmic Power and Politics* (New York: Oxford University Press, 2018), 69–72.

46 Matteo Pasquinelli, "How a Machine Learns and Fails – A Grammar of Error for Artificial Intelligence," *Spheres – Journal for Digital Cultures*, no. 5 (2019): 1–17.

47 Anthony Elliott, *The Culture of AI: Everyday Life and the Digital Revolution* (New York: Routledge, 2019), 107.

48 Jean-François Lyotard, *The Postmodern Condition* (Manchester: Manchester University Press, 1984), 44.

49 Patricia Owens, "Accidents Don't Just Happen: The Politics of High-Technology 'Humanitarian' War," *Millennium: Journal of International Studies* 32, no. 3 (2003): 616.

50 Elke Schwarz, "Autonomous Weapons Systems, Artificial Intelligence and the Problem of Meaningful Human Control," *Philosophical Journal of Conflict and Violence* V, no. 1 (2021).

51 Nobert Wiener, *The Human Use of Human Beings: Cybernetics and Society* (London: Free Association Books, 1989), 183.

REFERENCES

Anders, Günther, "Reflections on the H Bomb," *Dissent* 3, no. 2 (1956): 46–155.

Anders, Günther, *Die Antiquiertheit des Menschen: Über die Seele im Zeitalter der zweiten industriellen Revolution* (München: C.H. Beck Verlag, 2010).

Anduril Industries, "Anduril Raises $450 Million in Series D Funding," *Medium*, June 17, 2021, https://medium.com/anduril-blog/anduril-raises-450-million-in-series-d-funding-671f0a27876b.

Bhuta, Nehal, Susanne Beck, Robin Geiss, Hin-Yan Liu and Claus Kress (eds.), *Autonomous Weapons Systems: Law, Ethics, Policy* (Cambridge: Cambridge University Press, 2016).

Blodget, Henry, "Mark Zuckerberg, Moving Fast and Breaking Things," *Business Insider*, October 14, 2010, https://www.businessinsider.com/mark-zuckerberg-2010-10?r=US&IR=T.

Boslaugh, David L., *When Computers Went to Sea: The Digitization of the United States Navy* (Los Alamitos: IEEE Computer Society Press, 1999).

Brockman, John, "Introduction," in *Possible Minds: 25 Ways of Looking at AI*, ed. John Brockman (London: Penguin, 2019), xv–xxvi.

Bucher, Taina, *IF... THEN: Algorithmic Power and Politics* (New York: Oxford University Press, 2018).

Conger, Kate, and Cade Metz, "'I Could Solve Most of Your Problems': Eric Schmidt's Pentagon Offensive," *The New York Times*, May 2, 2020.

Congressional Research Service, "Defense Primer: U.S. Policy on Lethal Autonomous Weapons Systems," *Federation of American Scientists*, December 1, 2020, https://fas.org/sgp/crs/natsec/IF11150.pdf.

Dougherty, Chris, "Want an Agile Pentagon? Don't Go Chasing Waterfalls", *DefenseOne*, March 22, 2021, https://www.defenseone.com/ideas/2021/03/want-agile-pentagon-dont-go-chasing-waterfalls/172835/.

Elliott, Anthony, *The Culture of AI: Everyday Life and the Digital Revolution* (New York: Routledge, 2019).

Ellul, Jacques, *The Technological Society* (Toronto: Random House, 1964).

Feenberg, Andrew, *Transforming Technology: A Critical Theory Revisited* (New York: Oxford University Press, 1991).

Freedberg Jr., Sydney, "Army's New Aim is Decision Dominance," *Breaking Defense*, March 17, 2021, https://breakingdefense.com/2021/03/armys-new-aim-is-decision-dominance/.

Garson, David G., *Public Information Technology and E-governance: Managing the Virtual State* (Sudbury: Jones and Bartlett Publishers, 2006).

Heidegger, Martin, *The Question Concerning Technology and Other Essays* (New York: Harper & Row Publishers, 1977).

Heims, Steve J., "Introduction," in *The Human Use of Human Beings: Cybernetics and Society*, ed. Nobert Wiener (London: Free Association Books, 1989), xi–xxiii.

Holland Michel, Arthur, "Known Unknowns: Data Issues and Military Autonomous Systems," in ed. *United Nations Institute for Disarmament Research*, Geneva: United Nations Institute for Disarmament Research 2021, 1–41.

Knight, Will, "The Foundations of AI are Riddled With Errors," *WIRED*, March 31, 2021, https://www.wired.com/story/foundations-ai-riddled-errors/.

Knight, Will, "The Pentagon Inches toward Letting AI Control Weapons," *WIRED*, May 10, 2021, https://www.wired.com/story/pentagon-inches-toward-letting-ai-control-weapons/.

Luckey, Palmer [@PalmerLuckey], "We just raised $450M in Series D funding for Anduril," *Twitter*, June 18, 2021, https://twitter.com/PalmerLuckey/status/1405595273263423490.

Lye, Harry, "Will Roper on AI and autonomous in the US Air Force," *Airforce Technology*, December 3, 2019, https://www.airforce-technology.com/features/dr-will-roper-on-ai-and-autonomy-in-the-us-air-force/.

Lyotard, Jean-François, *The Postmodern Condition* (Manchester: Manchester University Press, 1984).

Marcuse, Herbert, "Some Social Implications of Modern Technology," *Studies in Philosophy and Social Science* 9, no. 3 (1941): 414–439.

Matthias, Andreas, "The Responsibility Gap: Ascribing Responsibility for the Actions of Learning Automata," *Ethics and Information Technology* 6, no. 3 (2004): 175–183.

McGee, Patrick, "Silicon Valley reboots its relationship with the US military," *Financial Times*, May 17, 2021, https://www.ft.com/content/541f0a02-ea27-43a4-b554-96048c40040d?shareType=nongift.

Metz, Cade, "Away from Silicon Valley, the Military is the Ideal Customer," *The New York Times*, February 26, 2021. https://www.nytimes.com/2021/02/26/technology/anduril-military-palmer-luckey.html.

National Security Commission on Artificial Intelligence (NSCAI), "Final Report," 2021. https://www.nscai.gov/

Owens, Patricia, "Accidents Don't Just Happen: The Politics of High-Technology 'Humanitarian' War," *Millennium: Journal of International Studies* 32, no. 3 (2003): 595–616.

Pasquinelli, Matteo, "How a Machine Learns and Fails – A Grammar of Error for Artificial Intelligence," *Spheres – Journal for Digital Cultures*, no. 5 (2019): 1–17.

ResearchAndMarkets.com, "Global Artificial Intelligence Military Market (2020–2025)," *Business-Wire*, March 23, 2021, https://www.businesswire.com/news/home/20210323005739/en/Global-Artificial-Intelligence-in-Military-Market-2020-to-2025---Incorporation-of-Quantum-Computing-in-AI-Presents-Opportunities---ResearchAndMarkets.com.

Ross, Casey, "Machine Learning is Booming in Medicine. It is Also Facing a Credibility Crisis," *Stat+*, June 2, 2021, https://www.statnews.com/2021/06/02/machine-learning-ai-methodology-research-flaws/.

Schwarz, Elke, "Autonomous Weapons Systems, Artificial Intelligence and the Problem of Meaningful Human Control," *Philosophical Journal of Conflict and Violence* V, no. 1 (2021): 53–72.

Shane, Scott, and Daisuke Wakabayashi, "'The Business of War': Google Employees Protest Work for the Pentagon," *The New York Times*, April 4, 2018, https://www.nytimes.com/2018/04/04/technology/google-letter-ceo-pentagon-project.html.

Sharkey, Noel, "Guidelines for the Human Control of Weapons Systems," *ICRAC Briefing Paper*, April 2018.

Sparrow, Robert, "Killer Robots," *Journal of Applied Philosophy* 24, no. 1 (2007): 62–77.

STM, "KARGU Rotary Wing Attack Drone Loitering Munition System," not dated, https://www.stm.com.tr/en/kargu-autonomous-tactical-multi-rotor-attack-uav.

Sullins, John P., "When is a Robot a Moral Agent?" *International Review of Information Ethics* 6, no. 12 (2006): 23–30.

UN Security Council, "Letter Dated 8 March 2021 from the Panel of Experts on Libya Established Pursuant to Resolution 1973(2011) Addressed to the President of the Security Council," March 8, 2021.

US Department of Defense, "Department of Defense Directive 3000.09," November 21, 2012, (Incorporating Change 1, May 8, 2017), https://www.esd.whs.mil/portals/54/documents/dd/issuances/dodd/300009p.pdf.

Wiener, Norbert, *The Human Use of Human Beings: Cybernetics and Society* (London: Free Association Books, 1989).

Winner, Langdon, "Three Paradoxes of the Information Age," in *Culture on the Brink: Ideologies of Technology*, ed. Gretchen Bender and Timothy Druckrey (Seattle: Bay Press, 1994), 191–204.

CHAPTER 7

A Roadmap for Living and Working with Intelligent Machines

Mark Fenwick and Erik P. M. Vermeulen

PART I – HELLO NEW WORLD OF ARTIFICIAL INTELLIGENCE (AI)

As university professors, summer has always been the best time to reflect on the content of our courses. We regularly update our teaching materials to keep them relevant. As educators, we should ensure that students are given the latest insights and trends; and teaching new issues also keeps us fresh and helps with our research and other writing. Recently, however something feels different. We have come to believe that a more fundamental change is necessary to meet the unique demands of a fast-changing technology-driven economy, culture, and society. Mere updates are no longer enough.

What is needed, instead, is a new approach to education, particularly in the fields of social sciences and law. To prepare the next generation of students for the future – for a world of intelligent machines – it has become necessary for everyone to think about and engage with cutting-edge developments in technology. We cannot avoid it. The alternative is to keep doing things the same old way and to risk drifting into obscurity.

But *where* should we start? *Where* are we now? We might take inspiration from Japan, for example. In March 2019, a group of experts argued that it is necessary to have all of the approximately 500,000 university and technical college students in the country take beginner-to-intermediate level courses on data science and AI.[1] They would thus be encouraged to acquire the skills necessary to apply AI solutions in their specialized fields of study.

We don't hear much from Japan these days (compared to the late 1980s when Japanese companies dominated the world). And yet, for various technologies, companies in Japan

DOI: 10.1201/9781003226406-8

are still global leaders. Interestingly, in the field of robotics, for instance, Japan adopts a different approach from other technology regions, such as Silicon Valley. The focus is less on "man versus machine," but on "man living in harmony with machine." For instance, Japan's robots come with stories surrounding them. This might be seen as childish or a gimmick. And yet, the goal is to smoothen the integration of robotic technologies into the everyday lives of the people who will have to work "with" them.[2]

This vision of seamlessly embedding intelligent machines into everyday life can offer a starting point for a reappraisal of education in a digital age. At least, it points us toward several important questions:

- What should we – as educators – be doing to prepare for a world where we have to co-exist with intelligent machines?

- How can we live in "harmony with these machines?"

- What should we be teaching students to ensure that they have the best chance of finding success – both in work and life – in the near and medium-term future?

The World Today

Before we prepare for the future, it might be helpful to consider the world as we experience it now. It seems clear that people increasingly feel lost and confused by the current direction of travel. There are several connected reasons for this sense of drift.

The Exponential Growth of New Technology

This is the single most significant difference between the world of today and the world of even three decades ago. Humanity has always experienced technological innovation – there is nothing new with that – but the speed of innovation and the fact that innovations appear to accelerate each other distinguishes the digital age from previous technological revolutions. Think of how big data, robotics, and AI are feeding off each other to drive various near-future technological developments, such as augmented and virtual reality, voice and mind interfacing, as well as quantum computing. It becomes evident that the process of the digitization of everything is ongoing and quickening, and that new forms of technology will continue to transform every aspect of our lives and the world. And as these technologies move into less obvious fields, the scope and scale of that disruption will become more far-reaching. #Digital is transforming and will continue to transform what it means to be human.

Authority Lost and a Creeping Sense of Crisis and Mistrust

These technological changes have had several disruptive social and cultural effects. Consider the following examples of such technology-driven social change:

- Traditional sources of knowledge and expertise are disrupted as trust in all forms of authority – most obviously, scientific and political authority – declines. It is replaced by, in the best case, highly contingent agreements (the "wisdom of the crowd," if you like) that can only be relied on for a short period before they become obsolete or irrelevant.

- The "old world" of curated news distributed by media conglomerates or autocratic governments are disrupted by new communication platforms and the viral reproduction of unmediated information.

- Well-defined roles and career paths in stable organizations and social systems are disrupted by more fluid identities in the continually evolving spaces and networks of a global, digital culture.

The scale of the social and cultural impact of digital technologies justifies talk of a digital "revolution." The effects of these changes are everywhere. We all now live in a state of profound technological and socio-cultural upheaval.

All of this new technology makes life more comfortable, more efficient, and creates new possibilities, but the acceleration in the development of technology is often experienced by individuals as threatening traditional ways of life and this becomes the source of tremendous uncertainty. People are anxious about the effects of technological change and where it is taking us. Concerns about the loss of privacy and control are the most visible manifestation of such anxieties. But, a sense of living with existential crisis has become the new normal, something that COVID-19 has merely served to reinforce.

The effect of this disruption is that traditional institutions within our society – corporations, mass media, and political organizations, for instance – are widely perceived as irrelevant and failing. They are not meeting expectations. And as these institutions lurch from crisis to crisis, this sense of failure grows, further feeding feelings of uncertainty and disorientation.

Everybody Is Now an Entrepreneur

The economy has similarly been transformed by digital technologies. The traditionally dominant, industrial corporations and their business models have been disrupted by the emergence of new businesses that leverage digital technologies (notably the combination of smartphones and the Internet) to offer previously unimagined (and impossible) products and services.[3] We are all familiar with the names: Amazon, Apple, Facebook, and Google have become some of the biggest companies in history. More generally, the economy has changed – the "gig" economy, the platform economy, the service (as opposed to product-driven) economy. There are plenty of labels vying to describe the newly emerging global economic order.[4]

Crucially, these macro-economic shifts also play out at an individual level. People find their jobs are replaced by machines or that interacting with machines occupies more and more of their work time. And more people worry about the possibility of their job disappearing. Such concerns are not confined to blue-collar work. Knowledge and other white-collar workers are similarly affected. Lawyers, for example, see many "simple" tasks being automated as RegTech becomes more sophisticated, and this leads many lawyers – and particularly younger lawyers who traditionally performed these routine tasks as part of their training – to wonder about what legal work will be left for the "lawyer of the future" as AI becomes more sophisticated.[5]

This concern is particularly pressing for the generation of students who are currently enrolled in law school. They may find it impossible to learn on the job, as more basic legal tasks are outsourced to machines, and the rationale for employing them at all is squeezed. The warnings of high-profile public figures from the worlds of technology and academia – think Elon Musk, Bill Gates or Stephen Hawking – concerning the disappearance of traditional jobs seem more real and pressing.[6]

As such, people are increasingly concerned with their economic prospects in a digital age. They feel that they can no longer rely on traditional institutions to provide any kind of financial security, even in the short-to-medium term. Equally, people expect more control and freedom to do their own thing and build a unique identity or brand. As we live longer, healthier lives, we see more, we experience richer, more diverse lives.

The result of these economic changes is that it becomes more important to know how to develop yourself in order to remain relevant in a digital economy. As traditional jobs and opportunities are disappearing, everyone needs to be more entrepreneurial.

A New Relationship to the Future

On a more abstract level, our relationship with the future has also changed. The hope that we might anticipate or otherwise predict the future from what we have learned from the past (either from history or from verifiable experience) has faded as a result of the new cognitive and normative uncertainties that characterize a digital age.[7]

Prediction becomes impossible in an age of exponential technological growth and technology-driven social and economic change. Things move too fast and in unknowable directions, leaving us incapable of even grasping the present let alone identifying a trajectory that can form the basis for any reliable predictions about where we might be going. In thinking about this open, undecidable future, we need to combine imagination with the dawning realization that such predictions are very likely to be wrong.

Writing about the future is less an exercise in genuine anticipation and more of a call for action. Not so much, "this is what will happen," and more "this is what I want to happen." Giving people the resources to construct and defend a credible vision of a possible future that can orient action in the present becomes a crucial function of education.

Forces Defining our Digital Future

In this spirit of predicting our digital futures, here are four trends – or forces – that we believe will continue to shape our collective tomorrow over coming decades.

The Digitization of Reality

Technology will continue to be integrated into every aspect of everyday life. And this includes everyday lived experience, as well as business and working experience. This much seems obvious. The emergence and proliferation of technologies mean that we already now live in a world of "ubiquitous computing" – a term that was first coined in the late 1980s by Mark Weiser, chief scientist at Xerox. Weiser imagined a future in which people interacted with and used computers without thinking about them or even being aware of them. On this image of the future, computers would "vanish into the background," weaving "themselves into the fabric of everyday life until they are indistinguishable from it."[8]

This proved a remarkably perceptive observation, and digital devices do indeed now form the invisible infrastructure of our daily lives and world. The Internet of Things further links the physical and digital worlds.[9] A global network of connected sensors records everything. The information that is generated by these interfaces can then be utilized for various purposes, including improving performance of diverse technologies. The information once converted to a digital form and managed and analyzed via software, predictive algorithms, and more sophisticated emergent AI creates a digital reality that defines and generates new forms of trust and truth.

Data-Driven Decision-Making

Data-driven decision-making refers to a process that involves collecting data based on measurable goals, analyzing patterns and facts from this data, and utilizing them to identify choices strategies that benefit a business or other organization across several areas.[10] Data-driven decision making involves leveraging verified, analyzed data as the basis of organizational choices rather than merely "shooting in the dark" or blindly following previous practice.

The massive proliferation of sensor-generated data and algorithms capable of analyzing such data facilitates faster and more accurate decision-making. Crucially, such decisions are increasingly made by machines, as it is only machines that are capable of handling the vast amounts of information generated. In this new digital world of evidence-based choices, automated algorithmic decisions are quicker, faster, and more precise.

Peer-to-Peer Culture, "Democratization" and "Disintermediation"

As our world becomes digitized and data-driven decision-making becomes more pervasive, many of the intermediaries who previously controlled and dealt with information become less relevant. We can see this disruption across many professions. Think of lawyers, bankers, consultants.

This process of disintermediation may extend to politics and may see more referenda; "direct" democracy in which traditional intermediaries – the legislative representatives of the people – are bypassed. Also, we can see more attention is given to encouraging active shareholders in companies and the use of technologies that facilitate a more significant role for shareholders in firm governance.[11]

The Centrality of AI for Everything

One last trend is the increasing pervasiveness of AI in our everyday world. Whatever you may think of new digital technologies, such as AI, blockchain, and the Internet of Things, it cannot be ignored that they are changing the way we live and work. It's happening all around us. Machines and intelligent systems are gradually taking over more and more routine tasks. For the moment, these machines and systems are mainly designed to provide assistance for specific tasks in specific situations. They augment our intelligence and improve our skills.

Consider the automotive industry. Collision warning, automatic braking, lane departure warning, and blind-spot notification systems are standard features on many of today's cars. AI and other digital systems do not only become more common, but they also become more accurate. For instance, AI already outperforms human lawyers in reviewing and annotating standardized contracts.[12]

And, there are, of course, plenty of other fascinating examples ranging from "chat-bots" in healthcare, data analytics in agriculture, blockchain technology in the energy sector, and robot-advisors in banking and finance. These examples make it very clear that there is no turning back. We have to get used and prepared for living and working with intelligent systems and machines.

Preparing for the future will only become more crucial as machines become more advanced, connected, and autonomous. We don't intend to make any predictions here. Still, we expect that the automation trend will only accelerate and that AI will become the new normal.

The reasons for this are simple and straightforward. The current digital technologies should not be looked at in isolation. The proliferation of Internet of Things devices and sensors will lead to the availability of vast amounts of data, which will lead to smarter and more autonomous systems. Clearly, we need AI to make sense of the data. Investments in AI-related technologies are snowballing (see Figure 7.1).

Corporations – including old-world manufacturing giants – are all now betting on new digital technologies. Mentions of the words digital and technology in earning calls are increasing. We see a similar trend in the acquisitions of digital technology start-ups.

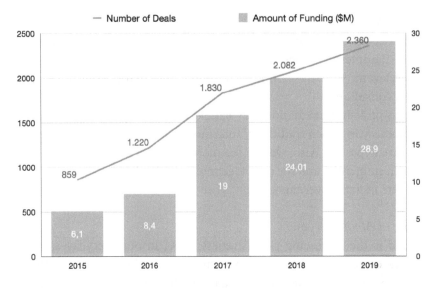

FIGURE 7.1   AI equity deals, 2015–2019.

Source: CB Insight.

Governments follow suit by starting projects to digitize processes, "eliminate paper," and embrace a "cashless society."

Consumer expectations and demands of digital technologies are also changing. The majority already use AI-powered devices of service without actually realizing and understanding. Yet, when new tech products are introduced, they are often disappointing. Expectations are simply higher. Faster, safer, and smarter services appear to be very attractive to consumers.[13] We believe in digital technologies and tend to focus on the many advantages that it brings today and will continue to bring in the future.

Computers, algorithms, and big data improves our health and help medical doctors make better diagnoses. Sensors, blockchain technology, and AI lead to more transparency and a level playing field in the global food supply chain. This results in more convenience – a frictionless user experience – as well as increased productivity. It also helps us be more creative (because we are less reliant on hierarchies and centralized systems). Finally, it facilitates financial inclusion in emerging economies.

And yet, we must also acknowledge that smart and connected AI-driven digital technologies will have numerous challenges that cannot be ignored. Large scale unemployment among both manual and knowledge workers is often mentioned. The same goes for social inequality as a result of automation. The concepts of privacy, data protection, and ownership all have to be re-imagined.[14] This will become even more important as the Internet of Things economy develops further. Machines and devices will act, interact, and transact more often as they share and use more data. The last thing anyone wants is for personal data to be shared with a "bad actor device." We, therefore, need to turn our attention to potential issues and threats without stifling socially valuable disruptive innovation.

What's Next?

These trends are going to transform every aspect of how we live and work. To take a legal example again, think of how contracts and contractual disputes will be disrupted by smart contracts. In this context, a smart contract refers to a computer program code or protocol that automates the verification, execution, and enforcement of specific terms and conditions of a contractual arrangement.[15] A smart contract can be defined as computer code that automatically executes all or parts of an agreement and is then stored on a blockchain-based platform. To take a simple example, consider a smart contract in the case of a car loan. If the borrower misses a payment (tracked via technology) then the contract/code would not allow the use and operation of that car, i.e., the contract would be "enforced" via networked technologies that disable the car, rather than a repo man. Such "contracts" may produce efficiency, timing, and performance improvements as a result of the automation of the contract's terms.

Crucially, this automation is achieved via the computer code, which controls the automated performance in the context of an Internet of Things environment where digital devices are inter-connected to a decentralized blockchain. Imagine the automobile accident of the future in which cars communicate directly with insurance companies in order to self-execute the terms based on "facts" that are "agreed" on between the vehicles and the sensor data automatically generated at the time of the accident.

What is the role of the lawyer in this new social order? The "smart contract" in such a case only exists in the code, and the parties communicate automatically. Of course, the contract will still need to be "drafted," but again, this will be done in code form and not words. Beyond that, the traditional functions of the lawyer, related to execution and enforcement, will disappear as intelligent machines communicate with each other. This does not mean that lawyers will be rendered obsolete. But they will need to be able to operate in multidisciplinary teams, communicating with the coders and designers who will all play a crucial role in deploying a new and different kind of product and experience.

To navigate this complex and new reality, we need to embrace a different and more open mindset that can identify opportunities and solve the unique problems of a digital world. The increased reliance on robotics, Big Data, and AI mean that all of our lives will be disrupted. But it also means that, as consumers, we will have more time and freedom. Technology has reached a stage where it is possible to create this new reality. We won't have to deal with many of the daily things that occupy us now – the "nonsense of everyday life" if you like. Instead, we need to build new knowledge and apply that knowledge in a never-ending process of transformation.

Recall the warnings of Bill Gates, Elon Musk, and Stephen Hawking. We have to think about robotics and automation now. Perhaps, by doing so, we may emerge in an even more robust and durable place. Only by embracing new realities will we avoid being taken over by the machines. Now that smarter and more advanced robots are nearly here, we are free to be smarter. And, maybe, even a little bit more human. As educators, we need to prepare students for this future. We need to encourage all students – and that includes law students, business students and students of other social science – to engage with new technology and to think about what it means for their lives and the different careers they wish to pursue.

PART II – SMARTER EDUCATION, SMARTER AI

A sense of dissatisfaction with the current state of university education has everything to do with the exponential growth of technology. As mentioned above, we are experiencing an ongoing process of the digitization of reality. This is the result of the global proliferation of new technologies. The future will be full of tremendous opportunities, but it will also be a world of immense uncertainty. Such uncertainty creates a considerable challenge for educators. With the current pace of innovation and shorter innovation cycles, it seems evident that new technologies are going to continue to transform every aspect of how we live and work. Constant technological disruption is the new normal. "Old World" concepts, models, paradigms, and ideas will no longer be relevant.

So, *what* should we be teaching our students today? Teaching has always tended to be backward-looking, and transmitting the settled knowledge of the past has been the starting point for our whole approach to education. For example, in the field of law, students have traditionally analyzed existing laws, regulations, and cases. The idea has been that if you understand historical developments, you would be able to solve future problems by applying old doctrines and precedents to any new fact situation. A similar logic can be seen in other fields. MBA programs, for example, employ the same approach in a business context, using a similar case-studies method.

The function and responsibility of the educator was to transmit this settled, old content. In a world of information asymmetries, the educator-student relationship was, by necessity, a hierarchical one. After all, the teacher had all the knowledge. But, this model seems ill-suited in a world of constant change. If the future is radically different from the present, it doesn't make much sense to focus on content that is likely to be irrelevant or worse (i.e., damaging). Moreover, the ready availability of information means that the informational advantage of the teacher is of much less significance. The result? Education needs to become much more forward-looking and skills-based.

How then do we prepare the next generation for dealing with unknown future problems? This is the question that we need to be asking. Merely "updating" the content of our programs or courses is no longer adequate.

Understanding the Technology

For a start, everyone is going to need a much better practical understanding of the technologies surrounding computers, communication networks, AI, and Big Data. For many of us, the underlying technologies that are driving social change remain a mystery, and that is a problem.

Practical technical knowledge also needs to be integrated into the many levels and fields of education. Coding and data analytics seem like a good starting point.[16] Indeed, when a student made the following observation in a class that we were teaching on business strategies and organizations, we thought it was the most obvious thing in the world: *"Artificial intelligence is poised to transform the way we live, work and learn. It will have a huge impact on the way we do and organize a business."*

Of course, the comment isn't exact, and the term "AI" has many sub-categories. But what was surprising was the reaction of other people in the room. What seemed to be a simple statement of fact was rejected – quite aggressively – by others: *"The media headlines are just feeding the hype around AI"* … *"There are a lot of misconceptions about AI"* … *"AI doesn't have anything to do with real "human" intelligence"* … *"The AI applications out there aren't particularly intelligent/smart"* … *"We are still in the very early stage of discovery"* … *"We should not waste too much time on AI yet."*

This experience made us think about the general environment in which we now operate as educators. We need to "level up" everyone's knowledge of AI. Much more attention needs to be paid to AI literacy to overcome the dismissive skepticism that currently surrounds much of the discussion, even among "digital natives." The issue is too important to limit the talks to computer science or engineering. Artificial intelligence must become a mandatory subject at every level of education in order for the general population to be better informed as to the current state of play.

But, we also need to think about other skills and capacities that are important in a world of unprecedented technology-driven social change. The focus should be on building skills that will assist the next generation in making the right decisions under conditions of extreme uncertainty, as well as building technical understanding. Here are some suggestions about the kind of issues that we believe are important:

- **Creative Thinking and Communicating Creative Solutions.** The next generation has to be able to think fast and "out of the box." Dynamic analysis of complex situations and the ability to communicate those solutions, in presentations or in video form, will be essential.

- **Entrepreneurship.** In the future, we will see flatter, more open and looser organizations and social platforms. It is, therefore, important that the next generation finds ways to become more self-productive and self-motivating, i.e., how to operate without a "boss" telling them what to do. As traditional concepts of a "career" become much less relevant, it will become increasingly important to build a unique personal brand by developing and then transmitting the right kind of story.

- **Teamwork.** More open organizations mean having to work in teams of strangers, often from diverse national or disciplinary backgrounds. The ability to work in a diverse group, continually adapting to a changing situation and working patterns, becomes vital.

- **Ethics.** Many of the problems of the future will be ethically complex. This seems particularly true in the context of robotics and AI.[17] Yet, all new technologies raise difficult ethical issues. Building the capacity of students to think about ethics seems another way that teachers can add value.

- **Interdisciplinary Learning.** Finally, we need to be open to interdisciplinary and multidisciplinary study, however strange it might initially seem. For example – and this is just our personal opinion – we think that more extensive knowledge of biology can help prepare the younger generation for the challenges of the future. Partly, this reflects our own preference for biology metaphors for understanding recent changes in the business world. We believe that metaphors involving the "environment" and "evolution" are helpful. But, this also reflects our belief that the next big wave of innovation is likely to be in the field of biotechnologies, and that knowledge of the field will be at a premium. But, even if this particular prediction turns out to be wrong, the exposure to multiple perspectives can only help in preparing the younger generation for an open and uncertain future.

How Should We Teach the Next Generation?

What teaching methods do we need to employ to be more effective as educators in the digital world? Of course, this is not a new question. Educators have always reflected on how to improve their performance, and there is a lot of discussion on this issue right now. But too much of the current debate seems overly simplistic. In particular, the current focus on "distance learning" or "online teaching" can appear particularly problematic as old paradigms are applied to very different situations.

The idea that putting everything online makes everything OK seems a little naïve. Of course, it might give certain groups access to information that they otherwise wouldn't have, and that is obviously a good thing. But, we also need to be aware of the risks of such an approach. In particular, it preserves the traditional teacher-student hierarchy and focuses on traditional content. The shift to remote teaching in the context of COVID-19 seems to confirm these concerns.

Instead, we need to be creating flatter, more inclusive learning environments – "Labs" – in which students are challenged to be more creative and entrepreneurial. Students must be forced to work in teams and think about possible scenarios with the associated challenges and solutions. In this way, the capacities relevant to a digital age can be cultivated.

But, perhaps the most persuasive argument for changing the approach to education is the expectations and demands of young people today. It seems evident that the "next" generation expects something different from education. The traditional approach simply bores them, and they switch off. The temptation to play with their phone – or simply to skip class altogether – is too high. *"Why should I go to class if I can get the same (or even better) information online?"*

After all, the university entrants of today were born into and entirely raised in a digital world. They belong to a culture that has no memory of a pre-Internet age. They are completely immersed in digital culture and all its "effortless" possibilities.

New technology is driving an important shift in our culture and society, and this is both important and real.[18] What we have learned is that the young generation is in love with technology and the endless possibilities such technologies offer. They want to learn more about emerging technologies, such as AI, machine learning, blockchain, and smart contracts.

And yet, perhaps the most crucial observation is that the students genuinely believe that new technologies have made existing and traditional models, policies, and practices obsolete. Every area of life has been disrupted by technology, and the students are intuitively aware of this fact and are at home in this fast-changing world. The students expect education to help them prepare for the new opportunities and challenges of the digital age.

Here are five "demands" of the Millennial generation that need to be met.

- *"Challenge Me ... and Let Me Challenge You."* Technology has always helped improve the way we teach. For instance, computers and the Internet have significantly increased access to information. What distinguishes the digital age is the instant and almost effortless possibilities of accessing this much larger pool of knowledge.

  Students are all hooked to their screens and instantly check out and verify topics – in real-time – that are discussed in class. Instead of passively accepting and memorizing facts, the Millennial generation wants independent confirmation of what it is they are being told. Healthy skepticism is their default attitude and response. Students want to be challenged by thinking about the future and how they can contribute to building a better society and environment. But, they also want to be able to challenge us about what we are telling them. The result? A more dynamic and interactive classroom experience for everyone.

- **"Don't 'Teach' Me ... Augment My Experience."** The current generation of students doesn't feel the need to become "textbook smart." They know that "facts" can be easily found online through their own independent action. In a world of open access to knowledge, it makes little sense to rely on the classroom as a forum for the transfer of knowledge.

  Instead, the students prefer to learn from the stories and experiences of others. These shared stories and experiences help them to augment their own experience by learning from the success and mistakes of others. This helps them avoid making the same mistakes as their "influencers." The capacity to weave facts into a larger narrative becomes the key. As such, the focus of the in-class experience needs to be on providing opportunities for constructing, sharing, and "selling" such stories. An irony of a technology-driven age is that story-telling has never become more important.

- **"Build My Capacity for Entrepreneurship."** The Millennial generation realizes that traditional career paths are not available anymore or, at least, enormous risks attach to such careers. Nor do these traditional role models interest or inspire them. Everyone understands that they need to be more entrepreneurial, and everyone wishes to be more entrepreneurial.

  This generation of students understands that jobs are and will become more fluid and flexible. Against the backdrop of such uncertainties, many want to explore the possibility of starting their own business. Also, they want to know what they should do to become part of more prominent platforms, such as Amazon, etc. As such, education might provide them with a platform for their own experiments in content creation and business design. Education needs to be about gaining experience that is useful in the real world and offering the opportunity to make mistakes (and understand the value of making mistakes), as well as build core and transferable capacities.

- **"Prepare Me for a 'Flatter' World."** The Millennial generation understands that the exponential growth of new technologies offers a more level playing field. In a digital age, old hierarchies, as well as traditional roles and backgrounds, matter much less than before. Market value is (and will become) more contingent on personal talent and skills. On this type of account, education must be about creativity and self-direction.

- **"Inspire Me... and Give Me Freedom and Responsibility."** New technologies have made it possible to work anytime and from any place. Technology is enormously empowering. What we have learned is that Millennials can come up with amazing and creative solutions, if they are given an inspiring task and the freedom and responsibility to complete that task in any way that they wish.

  To tap into their talent, it is crucial to offer students meaningful and challenging assignments. Students demand this combination of inspiration, freedom, and responsibility. The Millennial generation is quick to "switch off" if they do not feel engaged. A lack of interest can easily be mistaken for apathy. What has become clear already is that a more open and constructive way of teaching is necessary to be able to adapt the format and content of the course to continuously meet the wishes and demands of the Millennial generation.

When we talk about the "digital age," we tend to focus on new technologies. This is necessary, but Millennials are particularly interested in how these technologies are transforming our everyday lives, i.e., on the economic, cultural and social effects of such technologies. Unfortunately, so much of what passes for education today (and there are, of course, exceptions) fails to offer something meaningful to Millennials. This is where many of the problems and misunderstandings about education in a digital age begin. What we need to be doing is to better understand the current context and re-design our whole approach to education based on the needs of this new reality.

PART III – WE SHOULD ALL STUDY AI

AI has the potential to make us all smarter, healthier, and more entrepreneurial. Thus, delivering more AI-literate students will undoubtedly be beneficial to society. Nevertheless, the skeptics are still under the impression that AI remains a field of study that belongs to an elite group of "AI specialists." Undoubtedly, the world's best AI researchers are continually making hugely important breakthrough discoveries. But what the skeptical position ignores is the fact that we can already see more and more implementation and applications of AI in our daily lives. The rapid development of better algorithms and computing power in combination with the increasing availability and digitization of information leaves no area of our lives unaffected. Think about healthcare, HR, marketing, law, energy, agriculture, shopping. The fact that we (as consumers) are, often unknowingly, sharing more and more data will only improve the currently available applications allowing for further new use cases.

But the pervasiveness of these technologies is also matched by a lack of visibility. By and large, we are unaware of how much AI-related technologies already impede upon our lives. AI-related technologies form an invisible architecture that organizes and structures our experience of the world.

It is for this reason that everybody needs to have a basic understanding of the working of AI. It has already become a necessary precondition for understanding today's world – a new form of literacy, if you like, and a necessary precondition of responsible citizenship. It makes no sense to leave the people most affected by these applications uneducated about what is *already* happening around them.

But for AI to reach its full potential in a non-destructive form, non-AI experts must be involved in the development of the technology. This means acknowledging a distinction between the conceptual development of such technologies – in which literate non-AI experts can make a meaningful contribution and coding and programming, which will continue to be dominated – as it should – by computer engineering. Currently, however, it is mainly AI specialists that are dominating discussions regarding all aspects of this field. Given the importance of this issue, a more diverse range of people should be actively involved in discussions and development of new products and services. Such involvement will not only help identify different technical and ethical problems but allow us to find more innovative solutions and reveal new possible use cases.

In this way, improving knowledge about AI will facilitate non-AI workers in contributing – in a meaningful way – to debates on how to improve the performance of algorithms and to limit abuse. This is in everyone's best interests. We should not confuse the complexity of the technology and its effects with the view that any view or opinion is equally valid. And, as these technologies become increasingly sophisticated and beyond the understanding of even those individuals most intimately familiar with inner workings, such questions of responsible design will only become more pressing.[19]

Big Data and ever more sophisticated intelligent machines are going to play an increasingly crucial role in the activities of knowledge workers. It will continue to drive a revolution in how research is conducted; how customers are identified, products are advertised and uncovered; conflicts are solved; and, organizations are managed and coordinated. Data-driven technologies will contribute to identifying solutions to global challenges as defined in the United Nations Sustainable Development Goals related to poverty, inequality, climate change, environmental degradation, prosperity, and peace, and justice.

And more knowledge about how algorithms operate will make us all more sophisticated consumers of AI solutions. Different forms of citizenship and new identities will emerge. This is a necessary condition in preparing everyone for a world where we will co-exist with intelligent machines. It will also encourage us to take AI more seriously and focus on the soft skills that will become crucial in a word of AI: creativity, social skills, working in teams, and the ability to engage in a life-long process of self-learning and improvement. These are the things that make us human and which will continue to matter in a world where we are living and working with intelligent machines.

NOTES

1 Minako Yamashita, "Japan Aims to Produce 250,000 AI Experts a Year," *Nikkei Asian Review*, March 30, 2019, https://asia.nikkei.com/Economy/Japan-aims-to-produce-250-000-AI-experts-a-year.

2 Ashlee Vance, "Japan's Obsessive Robot Inventors are Creating the Future," *Bloomberg Businessweek*, October 26, 2016, https://www.bloomberg.com/features/2016-hello-world-japan/.

3 Mark Fenwick and Erik P. M. Vermeulen, "The New Firm," *European Organization Business Review* 16, no. 4 (2015): 595–623.

4 Diane Mulcahy, *The Gig Economy* (New York: Amacom, 2016).

5 Janos Barberis, Douglas W. Arner, and Ross P. Buckley (eds.), *The RegTech Book: The Financial Technology Handbook for Investors, Entrepreneurs and Visionaries in Regulation* (Chichester: Wiley, 2019).

6 Quincy Larson, "A Warning from Bill Gates, Elon Musk, and Stephen Hawking," *freeCodeCamp*, February 19, 2017, https://www.freecodecamp.org/news/bill-gates-and-elon-musk-just-warned-us-about-the-one-thing-politicians-are-too-scared-to-talk-8db9815fd398/.

7 See Niklas Luhmann, "Describing the Future," in idem, *Observations on Modernity*, trans. William Whobrey (Stanford: Stanford University Press, 1992), 63–74.

8 Mark Weiser, "The Computer for the 21st Century," *Scientific American* 265, no 3. (September 1991): 94–104.

9 Samuel Greengard, *The Internet of Things* (Cambridge, MA: MIT Press, 2015).

10 Simone Gressel, David Pauleen, and Nazim Taskin, *Management Decision-Making, Big Data and Analytics* (London: Sage, 2020).

11 Mark Fenwick, Wulf Kaal, and Erik P. M. Vermeulen, "The Unmediated and Tech-driven Corporate Governance of Today's Winning Companies," *New York University Journal of Law & Business* 16, no. 75 (2019): 76–121.

12 ArtificialLawyer, "LawGeex Hits 94% Accuracy in NDA Review vs 85% for Human Lawyers," February 26, 2018, https://www.artificiallawyer.com/2018/02/26/lawgeex-hits-94-accuracy-in-nda-review-vs-85-for-human-lawyers/.

13 Wolfgang Lehmacher, "5 Reasons Consumers Will Embrace Artificial Intelligence," *World Economic Forum*, January 4, 2018, https://www.weforum.org/agenda/2018/01/consumers-will-embrace-artificial-intelligence/.

14 Kai-Fu Lee, "The Real Threat of Artificial Intelligence," *The New York Times*, June 24, 2017, https://www.nytimes.com/2017/06/24/opinion/sunday/artificial-intelligence-economic-inequality.html.

15 Marcelo Corrales, Mark Fenwick, and Helena Haapio (eds.), *Legal Tech, Smart Contracts & Blockchain* (Singapore: Springer, 2019).

16 Mark Fenwick, Wulf Kaal, and Erik P. M. Vermeulen, "Legal Education in a Digital Age: Why Coding Matters for the Lawyer of the Future," in *Legal Tech & the New Sharing Economy*, ed. Marcelo Corrales Compagnucci, Nikolaus Forgó, Toshiyuki Kono, Shinto TeramotoErik P. M. Vermeulen Marcelo Corrales Compagnucci, Nikolaus Forgó, Toshiyuki Kono, Shinto Teramoto, and Erik P. M. Vermeulen (Singapore: Spinger, 2020), 103–122.

17 Julia Bossmann, "Top 9 Ethical Issues in Artificial Intelligence," *World Economic Forum*, October 21, 2016, https://www.weforum.org/agenda/2016/10/top-10-ethical-issues-in-artificial-intelligence/.

18 Ryan Jenkins, "The Complete Story of the Millennial Generation," *Inc.*, July 10, 2017, https://www.inc.com/ryan-jenkins/the-complete-story-of-the-millennial-generation.html.

19 For an account of how technology is increasingly beyond human understanding, see Samuel Arbesman, *Overcomplicated: Technology at the Limits of Comprehension* (New York: Portfolio Press, 2017).

REFERENCES

Arbesman, Samuel, *Overcomplicated: Technology at the Limits of Comprehension* (New York: Portfolio Press, 2017).

ArtificialLawyer, "LawGeex Hits 94% Accuracy in NDA Review vs 85% for Human Lawyers," February 26, 2018, https://www.artificiallawyer.com/2018/02/26/lawgeex-hits-94-accuracy-in-nda-review-vs-85-for-human-lawyers/.

Barberis, Janos, Douglas W. Arner, and Ross P. Buckley (eds.), *The RegTech Book: The Financial Technology Handbook for Investors, Entrepreneurs and Visionaries in Regulation* (Chichester: Wiley, 2019).

Bossmann, Julia, "Top 9 Ethical Issues in Artificial Intelligence," *World Economic Forum*, October 21, 2016, https://www.weforum.org/agenda/2016/10/top-10-ethical-issues-in-artificial-intelligence/.

Corrales, Marcelo, Mark Fenwick, and Helena Haapio (eds.), *Legal Tech, Smart Contracts & Blockchain* (Singapore: Springer, 2019).

Fenwick, Mark, and Erik P. M. Vermeulen, "The New Firm," *European Organization Business Review* 16, no. 4 (2015): 595–623.

Fenwick, Mark, Wulf Kaal, and Erik P. M. Vermeulen, "The Unmediated and Tech-driven Corporate Governance of Today's Winning Companies," *New York University Journal of Law & Business* 16, no. 75 (2019): 76–121.

Fenwick, Mark, Wulf Kaal, and Erik P. M. Vermeulen, "Legal Education in a Digital Age: Why Coding Matters for the Lawyer of the Future," in *Legal Tech & the New Sharing Economy*, ed. Marcelo Corrales Compagnucci, Nikolaus Forgó, Toshiyuki Kono, Shinto Teramoto, and Erik P. M. Vermeulen (Singapore: Spinger, 2020), 103–122.

Greengard, Samuel, *The Internet of Things* (Cambridge, MA: MIT Press, 2015).

Gressel, Simone, David Pauleen and Nazim Taskin, *Management Decision-Making, Big Data and Analytics* (London: Sage, 2020).

Jenkins, Ryan, "The Complete Story of the Millennial Generation," *Inc.*, July 10, 2017, https://www.inc.com/ryan-jenkins/the-complete-story-of-the-millennial-generation.html.

Larson, Quincy, "A Warning from Bill Gates, Elon Musk, and Stephen Hawking," *freeCodeCamp*, February 19, 2017. https://www.freecodecamp.org/news/bill-gates-and-elon-musk-just-warned-us-about-the-one-thing-politicians-are-too-scared-to-talk-8db9815fd398/.

CHAPTER 8

AI, Chatbots, and Transformations of the Self

Anthony Elliott

INTRODUCTION

Klaus Schwab, founder of the World Economic Forum, has provocatively argued that AI is the most transformative force operating throughout the world today. Schwab refers to this transformation, or reinvention of the world, as a "fourth industrial revolution." The first industrial revolution was steam-powered, the second was electrical, the third was the arrival of the computer age, and today's is the age of digital transformation. Schwab contends that today's AI revolution is "unlike anything humankind has experienced before." The AI revolution is all this, and much more. The "much more" is, quite simply, that of reinvention! AI is not so much an advancement of technology, but rather the *metamorphosis* of all technology. The increasingly ubiquitous spread of software algorithms, deep learning, advanced robotics, accelerating automation, and machine decision-making penetrates in every aspect of society and personal life. From personal virtual assistants and chatbots to self-driving vehicles and telerobotics, AI is now threaded into large tracts of everyday life, and increasingly reshapes society and the economy.

As the huge wave of AI breaks across the world, the possibilities for the reinvention of common public life and successful collaborative social action on a scale unprecedented in human history has come sharply into focus. The upsides here are many. AI is being used, for example, to track fish in the Great Barrier Reef, protect biodiversity in the Amazon and deter animals from entering endangered habitats using sensors. Research is now at an advanced stage for AI-powered microscopes to monitor plankton floating in the sea and for robots powered by AI to clean the oceans. The deployment of AI in the fight against global terrorism, through the international pooling of information and intelligence from supercomputers, is another case in point. But there are also massive high-consequence,

DOI: 10.1201/9781003226406-9

possibly existential, threats. From killer robots to lethal autonomous weapons technology to AI used by criminal organizations or rogue states, AI might potentially spell the end of humanity – a warning issued by such luminaries as the late Stephen Hawking, Bill Gates, and Elon Musk.

In many respects what we are seeing today is a new agenda, at least in terms of public policy. The traditional welfare state, for example, was very much based on providing remedies to the fallout from problems once they had happened – if you lose your job, the state will provide benefits until you find a new one. Today we live in a very different world. We live in a time where robots move boxes in factories as well as conduct shelf-auditing in supermarkets, and where complex algorithms complete tax returns and trade on financial markets. The consequences of our increasingly automated global world involve a shattering of political orthodoxies. Politicians and policy-makers have to be much more interventionist, and craft policy thinking to cope with the unexpected, unanticipated shifts stemming from the digital revolution. Another major consequence of the AI revolution is that our entire parameters for assessing risk and opportunity radically change. Complex systems such as machine learning and advanced robotics are characterized by unpredictability, uncertainty, zigzags and reversals. Policy attempts aimed at coping with complex technological systems often produce unintended consequences, revealing or generating other issues or problems. Other solutions or synergies, in turn, emerge in response to such interdependencies between adaptive systems.

Consider, for example, recent breakthroughs in chatbots. In 2018, Google released *Duplex*. Advertised as the next big advance in AI, Google's breakthrough featured a human-sounding robot having conversations with people who couldn't easily tell that they were talking to a robot. *Duplex* generated massive public interest by carrying out various everyday tasks: making phone calls on the user's behalf to schedule appointments, make reservations in restaurants and book holidays. This virtual personal assistant, portrayed as the most sophisticated talking robot to date, deployed natural speech patterns that included hesitations and a range of affirmations such as "mmm-hmm" and "er." Many people found it extremely difficult to distinguish *Duplex* from an ordinary phone call. The breakthrough was facilitated by advancements in automatic speech recognition, text-to-speech synthesis and sociological research on how people pace their conversations in everyday social contexts.

Sociology is, arguably, of especial significance for helping to grasp the transformative impacts of *Duplex*. This is because the discipline of sociology has long focused on *speech*, or more accurately *talk*, as an essential medium for the production and reproduction of day-to-day life. The late sociologist Diedre Boden wrote that human sociability is created through "talk, talk, talk and more talk."[1] Talking person-to-person is not only how we exchange information, but also how we carry out many tasks, such as ordering pizzas, booking plane tickets and confirming meetings. And why Duplex is so fascinating, why people became held in thrall (again) to Google, is because it reveals the potential of AI to increasingly subcontract to robots the many tasks of talk which we conduct in everyday life.

It is estimated that more than 60% of internet traffic is now generated by machine-to-machine, and person-to-machine, communication. IT advisory firm Gartner has predicted that by 2022 the average person will be having more conversations with robots than with their partner. These claims may seem the material of science fiction, but they spell significant change at the level of the self. In this chapter, I want to probe further into these interconnections between AI, chatbots, and transformations of the self.

CHATBOTS AND VIRTUAL PERSONAL ASSISTANTS: FROM FACE-TO-FACE TO DIGITALLY MEDIATED COMMUNICATION

Just as texting changed written communication, the advent of sophisticated talking bots looks set to change the way we communicate with each other. Talking person-to-person is not only how we have traditionally exchanged information but also how people have carried out many tasks of daily life, such as ordering pizzas, booking plane tickets and confirming meetings. In the aftermath of AI, especially advances in deep learning and natural language processing software, is these very tasks that we are today increasingly subcontracting to chatbots.

When we communicate face-to-face there is, as the late American sociologist Erving Goffman put it, an expectation of "mutual attentiveness."[2] But these norms could be wholly deconstructed if we were to have the majority of our conversations with intelligent machines. Unlike face-to-face talk, chatbots do not require us put effort into making the conversation polite or interesting. We don't need to be charming, amusing, or assert our intelligence. Bots don't need to like us, even if we have a need to be liked. In fact, this would wildly complicate matters. A machine will simply extract the information it needs to create an appropriate response. It is possible that talking to machines all the time could re-engineer the way we have conversations. We could end up with the linguistic equivalent of emojis. As an article in the *New York Times* recently put it, interacting with robots could "mean atrophy for our social muscles."[3] If they're just machines, why bother with pleasantries?

The scientific research on this is still unclear. Some studies have found that people can actually be remarkably cordial to robots, while other research suggests people are liable to be rude and curt when they know conversational partners are not human. Such behavior could bleed into everyday life. Tech companies are already trying to head off this problem. After fielding concerns from parents, Amazon created a politeness mode for its Echo devices that gently reminds its users to say "please." Some chatbots are being developed to go even further and mimic human emotion. For example, clinical psychologist Alison Darcy built a talking bot to help people with depression and anxiety. The delightfully named Woebot spoke to 50,000 people in its first week of deployment – more than a human psychologist could speak to in a lifetime. In a study with 70 young adults, it was found that after two weeks of interacting with the bot, the test subjects had lower incidences of depression and anxiety. Users of the Woebot were impressed, and even touched, by the software's attentiveness. As one interviewee said: "Woebot felt like a real person that showed concern."[4]

In 1950, the computer scientist Alan Turing designed an experiment to answer one of science's most enduring questions: Is it possible to create a robot that could be mistaken for a human? To date, this has generally been answered in the negative – although recent advances in machine learning have given some pause for thought on this. The reason

Turing's question has been answered in the negative is that AI devices respond to speech by drawing from an enormous database of code, scripted utterances and network conversation. So, they can rarely respond to the unexpected shifts in, and immense complexity of, human conversation, save in minor ways. Brian Christian notes of such machine talk: "What you get, the cobbling together of hundreds of thousands of prior conversations, is a kind of conversational purée. Made of human parts, but less than a human sum."[5]

Machine language no doubt is significantly different to everyday talk in various respects (and we will consider these differences in more detail shortly), but this does not mean that people can always distinguish differences easily between their online and offline worlds, between the culture of AI and everyday social life. One powerful account of why this is so comes from MIT psychologist Sherry Turkle. In her pioneering book *Alone Together*, Turkle connects the culture of AI and digital technologies to the erosion of emotional intelligence.[6] For Turkle, AI creates the illusion of connectedness – through Amazon recommendations, virtual personal assistants and robotic pets – but the self as a result becomes increasingly brittle and drained. "Our new devices," writes Turkle, "provide space for the emergence of a new stage of the self, split between the screen and the physical real, wired into existence through technology."[7] For Turkle, today's digital world of social media, computational gaming and sociable robots is impoverishing the self. The digital revolution places ever new demands on our emotional resources, as people have become increasingly "open to the idea of the biological as mechanical and the mechanical as biological." Especially at risk here is the emotional constitution of childhood itself. "Children," writes Turkle,

> need to be with other people to develop mutuality and empathy: interacting with a robot cannot teach these. Adults who have already learned to deal fluidly and easily with others and who choose to 'relax' with less demanding forms of social 'life' are less at risk.[8]

Digital technologies, in other words, warp the self out of emotional shape.

While Turkle's writings on the dangers of a new psychology of engagement are undeniably significant, I have elsewhere argued that this account is wanting.[9] One key problem is that the individual appears largely passive in relation to the digital world in Turkle's writings. Insufficient attention is devoted to the cultural context of digital technologies, especially the complex ways that people take up, appropriate and respond to communication and information in the digital field. The cultural forms through which individuals engage with digital technologies today are not a one-way, overwhelming or debilitating process. *Pace* Turkle, life today is not so much captured as constructed through the digital. Turkle's claim that when individuals today engage with robots or digital objects they imagine forms of companionship shorn of the emotional complexities of mature relationships is inadequate. The division here between digital and actual, online and offline is dualistic. Rather than search for an answer to these conundrums at the level of individual psychology, it is instead necessary to develop a broader sociology of the self in the face of the digital revolution.

At this point, let us consider once again chatbots. Chatbots, and our engagements with them, raise many of the key issues which Turkle's analysis underscores. Talking with a chatbot, arguably, does involve a new "psychology of engagement," one that is very different from talking with colleagues, friends or family. From this angle, chatbots highlight very well some of the key differences between machine language and human language. Here I want to return to the insights of Goffman – whose writings on impression management and the self are especially fruitful for rethinking the dynamics of human communication (and especially of talk) in the era of intelligent machines. Goffman's sociology focuses centrally upon the skilled performances of human agents and especially underscores the significance of talk in everyday life. For Goffman, the core focus of our attention in social interaction is toward the "presence of others," or what might be termed the "norms of co-presence." Goffman uncovered in his work a kind of *invisible dimension of social life*, where people invoke various procedures for engaging in social activity, monitoring the actions and activities of other people and also the responses of others to their own actions and activities. In face-to-face talk "when the eyes are joined," [10] wrote Goffman, people display attentiveness, demonstrate commitment and assess the sincerity of others. Participants work at their joint performances of conversing, and crucially commit themselves to remain present and attentive for the duration of the social interaction, which entails effectively managing not only the flow of small talk but also of ensuing silences in order to perform talk and maintain the norms of co-mingling.

Let us consider Goffman's account in some more detail. According to Goffman, the impression management of the self and team performances occurs within an "action framework" which involves certain cultural conventions and social assumptions pertaining to appropriate social behavior. Frameworks of action also involve the positioning of people, or the spacing of bodies, as well as the physical features of the actual setting (furniture, equipment, spatial design, and so on). An individual acting within this framework will to a large extent adapt their behavior to the norms or rules relevant to the particular setting, and display an awareness of identity which sustains face-to-face communication. The action framework, in other words, feeds into the "impression" which individuals seek to convey to others as well as what actually unfolds in public areas of social life. A crucial distinction advanced by Goffman for understanding what goes on in social interaction is that of "front" and "back" regions.[11] Front region action and behavior generally requires of individuals that they engage in strictly controlled self-monitoring, both the monitoring of their own conduct as well as that of others. This can involve particular attention to social cues and the responses of others, as well as strong emphasis on the professional impressions that individuals are seeking to cultivate. A back region, by contrast, often involves action and behavior which might well discredit the impressions that a person is seeking to project in front region encounters. In back regions, individuals tend to "lower their guard" and act in ways which are free of the stresses and strains of front region impression management.

In most areas of everyday life, the front-region behavior of individuals in companies and organizations contrast with back region behavior, where individuals do not have to be overly concerned about the impression they seek to cultivate. In some sectors of commercial life – for example, in restaurants – the distinction between front and back regions is reasonably well-defined and fixed. Where individuals work in restaurant kitchens, there are often swinging doors or glass partitions to separate staff and diners, and the passage between these areas is usually strictly regulated. This is equally true of the reception areas of many organizations, which serve effectively as a transition point for the management of front and back region behavior. Goffman identified such regional demarcations as essential to social interaction and the impression management of the self. The question of how communication media might reshape processes of social interaction is something that Goffman touched on throughout his own writings, but for the most part only in a provisional and partial fashion. This is hardly surprising, given that Goffman developed his sociological approach at the historical moment in which the rise of mass media was only coming to redefine the twentieth century. Nonetheless, the profusion of electronically mediated materials into the social interactions of daily life is addressed, at least in part, in some of Goffman's writings.[12] The contrast between, say, a TV newsreader's formal presentation of self (with front region props of jacket and tie) while wearing jeans (below the news desk, beyond the reach of the camera and thus "sealed off" as a back region) is illustrative.

Goffman's account of presentations of the self powerfully underscores the immense skill which human agents display when in dialogue, with a strong focus on the mutual coordination of social interaction. For Goffman, dialogue is less about *language* in the sense of a linguistic system than it is about *everyday talk* situated in context. Goffman connects talk with the myriad forms of day-to-day experience, and his writings highlight that everyday talk in settings of co-presence rarely occurs in a technically smooth or ordered way. There are, for example, an array of contingencies, accidents, or hesitations which infuse everyday talk: people break into (and often disrupt) conversational flows; turn-taking is not perfectly ordered; and, in general, the talk which makes up conversations is of a fragmented nature. As Goffman notes, everyday talk is at considerable remove from the kind of "perfect talk" presented on, for example, TV or radio news bulletins. And this is precisely why Goffman refers to the production of the self as a skilled performance.[13] What gives everyday talk its precision is the skill of human agents in navigating the complexities of social interaction in situations of co-presence.

Such a standpoint is arguably insightful for grasping some of the core differences between everyday talk and machine talk, even though Goffman's writings pre-date the digital revolution. Digital devices deploying natural language processing programs and AI technology are plainly quite divergent from the ordinary conversations of people. Machine talk, as I have emphasized, occurs as part of pre-programmed sequences built up through machine learning. As a result, machine talk to date at any rate can usually only respond to conversational contingencies in quite minor ways. Digital devices might be programmed to convey an impression of "immediate talk" geared to the needs of the user, but the production of machine talk is, in fact, drawn from an enormous database of code, scripted utterances and network conversation. For example, most chatbots and virtual personal assistants consist of programmed "appropriate replies" to even the most obscure conversations. This is underscored by Christian's argument that machine language is a kind of *conversational puree*, a recorded echo of billions of human conversations.

It is not just Google's *Duplex* that is involved in this societal shift of talk toward the *conversational puree*. The chatbot ecosphere, and the rise of virtual personal assistants, is now populated by many – including Amazon's Echo, Apple's Siri, Facebook Messenger's M as well as the Google Assistant. These conversational assistants, while quite basic in operation, conduct dialogue which is somewhat similar to human conversation. More advanced, human-like chatbots, however, are expected to represent the future of our digital lives. For example, Ray Kurzweil – Director of Engineering at Google, and founder of the Singularity movement – has developed a chatbot ("Danielle") which demonstrates high-level human conversational characteristics. Especially significant is the claim by Kurzweil that Google will assist in turning your own identity into a bot. This individual customization of bots is the result of recent advances in AI, which is based on the feeding of data directly into software. As Kurzweil argues, "You can actually create one with your own personality if you feed in your blog, that expresses your style and personality and ideas, and (the bot) will adopt those."[14]

Central to many hyperbolic claims about the social impacts of chatbots is the ideology of techno-optimism.[15] A powerful discourse has assembled across many sciences and the public sphere which projects how AI in general and chatbots in particular will develop comprehensive knowledge regarding everything we do. The techno-optimist claim, in effect, is that chatbots will know us better than we know ourselves. Kurzweil is more cautionary, noting that chatbots are not yet capable of achieving the normal patterns of sociality realized in face-to-face talk. But he does argue that society is not far away from such a transformation, identifying 2029 as the year which will witness AI-powered human-quality machine talk.

The possibility of AI, software-driven chatbots demonstrating fully human-quality talk raises the questions of multiple complex futures, new kinds of social and system interdependence, as well as long-term and large-scale shifts in the nature of conversations and the links of those conversing – not only person-to-person but person-to-machine. I cannot pursue the sociological implications of these shifts in complex systems of automated talk here, although this is an issue I have discussed at length elsewhere.[16] My focus here concerns, instead, the impacts of automated talk on the self. The growing global trend to supplement social interaction with simultaneous digital communication has been experienced by many as socially alienating or fragmenting; but, equally, many enter routinely into such communicational multitasking. Unlike face-to-face interaction, in which the coordinates of space and time match for conversational participants, digitally mediated communication (including machine-to-person talk) involves complex intersections of time and space which are spliced or fused together by participants deploying digital technologies. This has involved a transformed space-time environment in which the significance of place has rapidly altered, and oftentimes largely diminished. Characteristic of much social life today is a novel blend of physical, communicative, digital and virtual interaction. Common in many households in developed countries today, for example, is engagement with multiple tasks on multiple screens – and often undertaken alongside of periodic face-to-face interaction. Whether using a smartphone, conducting a Skype meeting on an iPad or talking with a chatbot, more and more people split their attention across multiple tasks and multiple screens. These communications can be at once online and offline, with others physically present and those at-a-distance. All of these transformations impact deeply upon the self.

Also, increasingly characteristic of digital technologies in contemporary times is a related feature which again refers back to Goffman: the intrusion of back region events into front regions as well as everyday consciousness. The distinction between front and back regions, as previously highlighted, is rarely fixed. Goffman's work brilliantly chronicles the many ways in which back region behavior can "leak into" the self-presentation individuals seek to project in front regions. In conditions of digital life, however, there has been a multiplication of the interactive frameworks which redistribute relations between front and back regions. The injection of people's mobile phone use into public space is illustrative of this. Many conventional methods of impression management have largely declined in significance as a result of smartphone use in the public realm, such as when queuing at retail shops or waiting in airport terminals. Such situations involve a blurring of boundaries between front and back regions and thus often involve a collision of physical and digital interactive frameworks. This blurring refers to the changing character of the line between public and private life, as sometimes intimate and occasionally embarrassingly personal phone conversations are revealed for others to hear. But it is not just that digital technologies promote a merging of older and newer techniques of impression management, or the resorting of front and back region behavior. Rather, much front region behavior does not conform either to social expectation or cultural convention. Experimentation, and oftentimes improvisation, is actually the key to the character of contemporary digital life. In some cases, the back regions of digitally mediated interaction may simply be located around the edges of front region encounters. For example, a woman applying make-up while traveling on a train, preparing for an impromptu video conference call, is illustrative of this "melting" of front and back regions. Such developments promote a direct engagement with social practices which previously were sealed off as inappropriate for public display, or in Goffman's idiom only appropriate to back region behavior. Again, this represents an important change as regards the production and performance of the self.

CHATBOTS AND CONVERSATIONS: WHEN THINGS GO WRONG

We can best grasp the differences between day-to-day talk and automated machine conversation when something goes awry or there is a breakdown in technical functioning of the latter. As an instance of this phenomenon, glitches generated by Amazon's Alexa in accidentally recording and relaying human conversations can be mentioned. In 2018, it was widely reported that a family in Portland, Oregon, had received a phone call from an acquaintance advising them to disconnect their Amazon Alexa device. The reason? The device had unknowingly recorded private conversations in the family home and forwarded these, apparently randomly, to a person in the family's contact list. The conversation recorded, it transpired, was of a mundane nature – though commentators were quick to forecast the arrival of a dystopic world where chatbots spy on us.

The relatively perfect speech performance of chatbots and virtual personal assistants are clearly distinct from the complexities of ordinary day-to-day talk. Notwithstanding the immense technological advancements of natural language processing software such as Google's *Duplex*, everyday talk in face-to-face situations is, by contrast, much less ordered than that which is realized through AI-powered technologies. The managing of turn-taking in conversations rarely happens in such a way that people finish the sentences they are speaking, but this in itself indicates the immense skill and practiced learning that participants demonstrate in talking and listening to each other in situations of co-presence. And again, it is a key reason why machine language which is linear and pre-programmed so often falls short of the overall character of ordinary communicative exchange. Consider, for example, this statement which Amazon gave as an explanation of how snippets of the Portland family's private conversation were recorded by Amazon's Alexa and forwarded to a contact in their address book:

> Echo woke up due to a word in background conversation sounding like "Alexa." Then, the subsequent conversation was heard as a "send message" request. At which point, Alexa said out loud "To whom?" At which point, the background conversation was interpreted as a name in the customer's contact list. Alexa then asked out loud, "[contact name], right?" Alexa then interpreted background conversation as "right."[17]

Many commentators have argued that meeting the asymmetrical, inconsistent and spontaneous requirements of ordinary talk in a technologically error-free way is the next major challenge for AI natural language processing. But whether or not this is technologically feasible is not the point I am making. My argument is that what gives ordinary language its precision, as Wittgenstein showed, is its use in context - and this is something which, for the moment at least, sharply differentiates day-to-day talk and machine language.[18] As Christian writes, chatbots appear

> so impressive on basic factual questions ("What's the capital of France?" "Paris is the capital of France") and pop culture (trivia, jokes, and song lyric sing-alongs) – the things to which there is a right answer independent of the speaker. No number of cooks can spoil the broth. But ask it about the city it lives in, and you get a pastiche of thousands of people talking about thousands of places. You find it out not so much by realizing that you aren't talking with a *human* as by realizing that you aren't talking with *a* human.[19]

Even though we might be having less of them, human conversations are clearly not going to decrease in significance anytime soon. Nevertheless, the ubiquity of the smartphone has essentially liquefied our social world, which almost always includes a level of digital engagement with others outside the immediate social context. This has created a complex, contradictory mix of being present with others, even when they are not physically present. AI is not about the future – our lives are already saturated in it. Chatbots, softbots, and virtual personal assistants are becoming an integral part of our daily lives and our identities, even if we are not always aware of their role. If talking to chatbots and virtual personal assistants becomes the new normal, we should be aware of the ways they could change how we talk to each other, and how we relate to ourselves. One thing is certain. AI is having a profound impact on experiences of the self, what identity means, and of how selfhood intersects with others (both human and non-human) in the wider world.

NOTES

1 Deirdre Boden, *The Business of Talk: Organizations in Action* (London: Polity Press, 1994), 82 and 94.

2 Erving Goffman, *Encounters: Two Studies in the Sociology of Interaction* (Indianapolis: Bobbs-Merrill, 1961).

3 Clive Thompson, "May AI Help You: Intelligent Chatbots Could Automate Away All of Our Commercial Interactions – For Better or Worse," *The New York Times Magazine*, Tech and Design Issue, November 14, 2018, https://www.nytimes.com/interactive/2018/11/14/magazine/tech-design-ai-chatbot.html.

4 Clive Thompson, "May AI Help You: Intelligent Chatbots Could Automate Away All of Our Commercial Interactions – For Better or Worse," *The New York Times Magazine*, Tech and Design Issue, November 14, 2018, https://www.nytimes.com/interactive/2018/11/14/magazine/tech-design-ai-chatbot.html.

5 Brian Christian, *The Most Human Human: What Artificial Intelligence Teaches Us About Being Alive* (New York: Anchor Books, 2015), 25.

6 Sherry Turkle, *Alone Together: Why We Expect More from Technology and Less from Each Other* (New York: Basic Books, 2011).

7 Sherry Turkle, *Alone Together: Why We Expect More from Technology and Less from Each Other* (New York: Basic Books, 2011), 16.

8 Sherry Turkle, *Alone Together: Why We Expect More from Technology and Less from Each Other* (New York: Basic Books, 2011), 16.

9 Anthony Elliott, *The Culture of AI* (New York and London: Routledge, 2019).

10 Erving Goffman, *Behaviour in Public Places* (New York: Free Press, 1963), 92.

11 Erving Goffman, *The Presentation of Self in Everyday Life* (Harmondsworth: Penguin, 1959).

12 Joshua Meyrowitz, *No Sense of Place: The Impact of Electronic Media on Social Behavior* (Oxford: Oxford University Press, 1985).

13 Erving Goffman, *The Presentation of Self in Everyday Life* (Harmondsworth: Penguin, 1959).

14 Ray Kurzweil in Ethan Baron, "Google will Let You Turn Yourself into a Bot, Ray Kurzweil Says," May 31, 2016, viewed October 20, 2016, http://www.siliconbeat.com/2016/05/31/google-chat-bot-coming-year-renowned-inventor-says/.

15 For an informative discussion of 'techno-optimism', it is useful to consult: Katherine Dentzman, Ryan Gunderson, and Raymond Jussaume, "Techno-optimism as a Barrier to Overcoming Herbicide Resistance: Comparing Farmer Perceptions of the Future Potential of Herbicides," *Journal of Rural Studies* 48 (2016): 22–32.

16 Anthony Elliott, *The Culture of AI* (New York and London: Routledge, 2019).

17 Jason Del Ray, "Here's Amazon's Explanation for the Alexa Eavesdropping Scandal," *Recode*, May 24, 2018, https://www.vox.com/2018/5/24/17391480/amazon-alexa-woman-secret-recording-echo-explanation.

18 Ludwig Wittgenstein, *Philosophical Investigations* (Oxford: Blackwell, 1953).

19 Brian Christian, *The Most Human Human: What Artificial Intelligence Teaches Us About Being Alive* (New York: Anchor Books, 2015), 25–26.

REFERENCES

Baron, Ethan, "Google will Let You Turn Yourself into a Bot, Ray Kurzweil Says," May 31, 2016, https://www.siliconbeat.com/2016/05/31/google-chat-bot-coming-year-renowned-inventor-says/.

Boden, Deirdre, *The Business of Talk: Organizations in Action* (London: Polity Press, 1994).

Christian, Brian, *The Most Human Human: What Artificial Intelligence Teaches Us About Being Alive* (New York: Anchor Books, 2015).

Dentzman, Katherine, Ryan Gunderson, and Raymond Jussaume, "Techno-optimism as a Barrier to Overcoming Herbicide Resistance: Comparing Farmer Perceptions of the Future Potential of Herbicides," *Journal of Rural Studies* 48 (2016): 22–32.

Elliott, Anthony, *The Culture of AI* (New York and London: Routledge, 2019).

Goffman, Erving, *The Presentation of Self in Everyday Life* (Harmondsworth: Penguin, 1959).

Goffman, Erving, *Encounters: Two Studies in the Sociology of Interaction* (Indianapolis: Bobbs-Merrill, 1961).

Goffman, Erving, *Behaviour in Public Places* (New York: Free Press, 1963).

Meyrowitz, Joshua, *No Sense of Place: The Impact of Electronic Media on Social Behavior* (Oxford: Oxford University Press, 1985).

Ray, Jason del "Here's Amazon's Explanation for the Alexa Eavesdropping Scandal," *Recode*, May 24, 2018, https://www.vox.com/2018/5/24/17391480/amazon-alexa-woman-secret-recording-echo-explanation.

Thompson, Clive, "May AI Help You: Intelligent Chatbots Could Automate Away All of Our Commercial Interactions – For Better or Worse," *The New York Times Magazine*, Tech and Design Issue, November 14, 2018, https://www.nytimes.com/interactive/2018/11/14/magazine/tech-design-ai-chatbot.html.

Turkle, Sherry, *Alone Together: Why We Expect More from Technology and Less from Each Other* (New York: Basic Books, 2011).

Wittgenstein, Ludwig, *Philosophical Investigations* (Oxford: Blackwell, 1953).

CHAPTER 9

Computational Power in the Digital World

Massimo Durante

ARTIFICIAL INTELLIGENCE

Our reflection on artificial intelligence (AI) starts from the consideration of the lesson of a classic: Wolfgang Goethe's *Faust*. Let us briefly listen to Goethe, when Faust meditating on the translation of the beginning of the Gospel of John asserts: "It says: In the beginning was the Word." Not satisfied with the translation, he says again:

> It says: In the beginning was the Mind [Sinn]. Ponder that line, wait and see, lest you should write too hastily. Is mind the all-creating source? It ought to say: In the beginning there was Force [Kraft]. Yet something warns me as I grasp the pen, that my translation must be changed again. The spirit helps me! Now it is exact. I write: In the beginning was the Act [Tat].[1]

In Goethe's poem there is a straightforward progression from word (logos) to action through mind and force. Therefore, action is placed even before and above force and replaces thought and speech, which are normally understood as the vehicle and the expression of human intelligence. It is action that is placed at the beginning of all things. We are only at the beginning of the nineteenth century but in Goethe's work the perception that the world is not dominated by the mind but nor by force is clear: it is dominated by action. Word and mind indicate the existence of a sense and plan, namely, of a world ordered around some shared meaning, but also of a distance between the world and us, between our representation of things and their occurrence. Force shows in turn the existence of a resistance, that is, of an interaction with the world and with others. What does it mean or imply, for Goethe, to envisage that action is at the beginning of all things? Placed at the origin of all things, action finds its origin in itself: it is totally free. It is no longer the result

DOI: 10.1201/9781003226406-10

neither of a project, of an arrangement of things, nor of a resistance to be overcome: it is not even subject to a representation or a judgment. It is thus the action itself that creates and enacts the criteria based on which to represent, understand, and judge the world (this is a question that will form the subject of both Friedrich Nietzsche's and Carl Schmitt's meditation on the enactment of values[2]). For Goethe, when the gap between word and action is deleted and there is no longer a distance through which word can come to evaluate and criticize action, then everything becomes possible and the space of morality is reduced. However, this is not the context to devote to the exegesis of Goethe's Faust. What I want to observe is that for a long time there has been a progressive transition to action, to the stance placed on the ability to act, which nowadays culminates precisely, in an apparently paradoxical way, in the current development of AI systems.

In a sense, the roads of artificial intelligence and human intelligence are destined to increasingly move apart, in the years to come. This will not happen because, as feared by many, AI will be largely removed from the control of human intelligence but perhaps for a less obvious but certainly more relevant reason. AI will be less and less defined as the ability, typical of the human mind, to represent and understand reality; on the contrary, it will be more and more conceived as the ability to act and change that reality. As has been acutely observed:

> Artificial intelligence is any technology that we develop and use to perform tasks that would be defined as intelligent if they were created by a human being. This definition belongs to a proposal written in 1955 for a summer research project on artificial intelligence. [...] It is a counterfactual definition, since the capacity of artificial intelligence is defined as intelligent not in itself but when a human being would be defined as intelligent if he were able to achieve the same result. Today, artificial intelligence could be more efficient and more capable of performing a certain task – like winning a GO game – than any human being. This means that human players should be really smart to be just as good. Artificial intelligence is not and above all it does not need to be. We have outlined so far what artificial intelligence does; it is important to specify now what it is: it is a growing resource of *ability to act* which is interactive, autonomous and learns from itself. It is not necessary to consider artificial intelligence as 'intelligent', 'conscious' or 'similar to life', to pose serious problems to society.[3]

Against this backdrop, the history of AI is therefore the story of a progressive, unstoppable separation of intelligence from action. This separation raises three main consequences that we have to take into account when dealing with computational power, which is at the basis of AI systems.

First, we increasingly separate the ability to act from that of being intelligent: AI no longer needs to be strictly related to intelligence. It turns into "artificial agency."[4] This expression refers to AI systems' ability to act as well as to their ability to embody values in their own functioning.[5]

Second, artificial agency brings about two more consequences: on the one hand, we face the results of actions for which we do not always have a model of understanding and explanation,[6] on the other hand, AI systems may be structured *by design*[7] in such a way as to embody technological norms: that is, technologies are capable of conditioning human behaviors, enabling, facilitating or preventing them (*technological normativity*);[8] in this way, they implicitly incorporate norms in their functioning which are removed from discussion, deliberation and interpretation.

Third, as already observed, if the gap between word (logos or intelligence) and action is reduced, there is almost no room for moral discourse: in the realm of pure action, morality hardly becomes possible. Any moral evaluation risks drowning in the optimization of an action. In this sense, the trait that will characterize our age mostly will be the growing difficulty in distinguishing a correct answer from a successful action.

In this perspective, artificial agency may severely endanger human freedom from three different standpoints: empirical; epistemic; and normative. Actually, we might no longer be at the beginning of (1) a *chain of events*: in this sense, action is no longer only a human prerogative (empirical standpoint); (2) the *explanation of a chain of events*: we do not always have an adequate model to fully predict or retrospectively explain the effects of actions we are subjected to (epistemic standpoint); (3) the *responsibility for a chain of events*: we are not always able to attribute responsibility for an action, namely, to call an agent upon to account for its own action (legal/moral standpoint).

These problematic aspects of AI systems are made more acute and urgent by the progressive growth of the computational power that empowers these systems. For this reason, it is necessary to specify below what are the salient features of this computational power.

COMPUTATIONAL POWER

Our age is characterized by the impact of ICT (Information and Communication Technology) and by the rapid pace at which technologies evolve. As we experience and analyze the ongoing information revolution,[9] we are already witnessing the gradual shift from the so-called information societies to the current data-driven societies.[10] The growth of computational power (i.e., the ability to process inputs and to transform them into outputs) has accelerated the convergence between data and AI systems and has favored their development. As it has been remarked, the increase and diffusion of computational power is a driving force of our age.[11]

Hence, computational power feeds an ever-growing number of artificial agents, AI systems, machine or deep learning systems, or computational models, which are increasingly developed and exploited by big tech companies. This raises two related issues: (1) an epistemic issue: related to the specific understanding and representation of the world that is peculiar and instrumental to computational models: these models function based on their computational logic that is different from (and sometimes irreducible to) our own; (2) a normative issue: related to the use and the governance of computational power: this power is likely to be unequally distributed and exploited in society and this raises problems of control and governance over the modes and consequences of its deployment.

Knowledge and power are linked together in a new and peculiar way such that their re-arrangement is nowadays the decisive lever for the understanding and governance of our data-driven society. Let us focus primarily on the epistemic issue and afterwards on the normative question of digital governance.

We increasingly delegate tasks and decisions to computational systems, because of their efficacy and efficiency, but each of them has its own way of doing things, which depends on its unique knowledge and representation of the world as well as on its computing capacity. At times, it is even hard to speak of knowledge and representation in the same way as we generally understand them, because these computational systems operate based on epistemic models that are peculiar to them. Computational systems decide and act on the basis of their own knowledge and representation of the world, which is different from – and often irreducible to – ours. The spread of such systems hence produces a proliferation of new epistemic models and points of view, i.e., the number of representations of the world continues to increase, to the extent to which we have a (technological, economic, and social) interest in developing and implementing these systems.

But which representations of the world are going to prevail, and which ones will need to adapt? If for example we have an interest in developing the market for driverless cars, will humans be adapting to them or vice versa? Will we adapt the unstructured urban environment to driverless cars, perhaps by building preferential or reserved lanes? Or will we ensure that driverless cars adapt to our environment, by providing them with increasingly sophisticated AI systems? Will we develop a system that allows human beings to collaborate with driverless cars, letting them intervene in the driving process? Or maybe even one that allows driverless cars to collaborate with each other?

My hypothesis – which I have expounded elsewhere[12] – is that there will be a strong tendency to adapt ourselves and the world to the representation of that reality which is instrumental to the functioning of such computational systems. And this can happen not only because of economic and political reasons related to technological innovation – on which the growth of individual and collective well-being increasingly depends – but because of the very nature of computational power. This power, in fact, is not understood just as a sum of resources or means but rather as the ability to build a vision of reality and the world, based on computation (i.e., on the ability to process data and produce outputs), which creates the premises and the need for the development of such computational systems: in a word, which creates the conditions of its own enactment. However, the revolution and transformation of the world, engendered by the rise and spread of computational power, does not necessarily have to be seen as anything striking, subversive or disruptive. No, the change and revolution based on computational power is chiefly an "everyday revolution."[13]

It is therefore that much more profound, unnoticed and widespread, for it affects our customary habits and routines and alters the very texture of our day-to-day lives. Often without realizing it, we rely on computational systems to carry out an increasing number of daily tasks, which we progressively embody[14] into our forms of life: "One way of describing the direction in which our own culture is moving is that many of us are starting to *adapt* to what we might call a *digital form of life* – one which takes life in the infosphere for granted, precisely because the digital is so seamlessly integrated into our lives."[15] Of course, all this is not without consequences. Like any power, computational power falls into someone's hands. Hence, whoever owns and exploits computational power not only alters and directs the distribution of power (resources and means) that exists within a society but also favors and fosters the everyday revolution that is based on computational systems.

Computational power creates new ways of processing data and spreading information, anticipating the future, representing reality and how we act in it, even representing who we are and what we want. It essentially produces new forms of knowledge. This production of knowledge concerns not only the accrual of the knowledge-base and the training dataset that computational systems need to fulfill their tasks (their peculiar representation of reality), but also the representation that these computational systems present of ourselves (their peculiar representation of us). If the algorithmic procedure of a computational system comes up with a profile of a "good borrower," individuals will become also an object of such a representation.

The production of knowledge, based on the ability of computational systems to collect, aggregate, analyze data and infer new data from these, raises consequences that affect our daily lives. On the basis of the inferred data, for instance, abstract profiles of various kinds can be developed, according to which prerogatives, opportunities or rights are granted or denied (i.e., the selection of a CV, the granting of a loan, and the offer of a discount). Furthermore, let us remark that we are delegating not only practical tasks but also epistemic decisions: algorithms carry out research on key terms, select the news we consume, guide our consumer choices, facilitate our memory – or help us forget, outline our digital profiles, help us tell the story of who we are, reinforce our beliefs, deliver intelligent solutions, reinforce inaccurate perceptions, or, more simply, lead us up a garden path. As we have observed, not only are we confronted with new ways of understanding and representing reality, but we are also the object of this understanding and representation.

Is it possible, or desirable, to regulate these new forms of knowledge, which tend to regulate our own lives? Obviously, putting the problem in these terms would be a mistake. The production of knowledge based on computational power is an integral part of an important and in some respects essential process of technological innovation, and it also represents an advancement of knowledge. It is unthinkable to censure the development of artificial intelligence, algorithms and computational models as such. We should always resist the temptation to regulate, if not hinder, the proliferation of new forms of knowledge. We can and certainly have to regulate the contexts, tools and purposes of computational power, as well as the effects deriving from its increasingly widespread and pervasive use. This requires a difficult task, namely, to regulate the actors who mostly possess and exploit such power: this raises crucial issues related to the governance of computational power, since complex governance decisions are nowadays likely to generate dangerous side effects.[16]

DIGITAL GOVERNANCE

A chief aspect emerges in the field of digital governance, which has already been remarked acutely by Jack Balkin with regards to a specific domain, i.e., that of constitutional freedom of speech and expression of thought.[17] We have moved from a dyadic model of governance, based on the dialectic between rulers and ruled, to a triadic, if not even pluralistic, model of digital governance. The latter is based on the multiplicity of possible relationships between *public actors* (mostly states but also other supranational bodies), *private actors* (mostly digital infrastructure owners such as social platforms, media companies or search engines) and *individuals* (mostly network users that use digital services and infrastructures in a countless number of everyday tasks and activities, while producing and sharing online content).[18]

In the analogical world, the governance structure was essentially dual and consisted of the relationships between public and private actors. Of course, relationships were manifold and the subjects of relationships changed. But what remained identical was our tendency to interpret and decipher these relationships as dual conflicts, that is to say, as power oppositions between two actors at stake (this was valid both in the case in which the actors were historically determined subjects, as in the class struggle; and in the case of a conflict between metahistorical forces, as in the contrast between state and market, culture and nature, or capitalism and the environment). These relationships have been consistently understood as a "game with two players,"[19] to use an expression made famous by Michel Serres. The proliferation of the actors involved has multiplied the number of relationships at stake and marked the transition to a triadic or pluralistic model of governance that is much more complex than the previous one was. The new digital governance model is further complicated by the fact that we continue to examine its constitutive relationships as a set of two-player games or dual conflicts.[20]

Against this backdrop, it must be said that none of the actors involved has, at present, a clear digital governance model: (1) public political actors find it difficult to figure out an adequate model of governance for a global, dynamic, technologically designed reality, based on traditional political and legal categories that do not always suit a constantly changing world; (2) private economic actors (e.g., big tech companies) have a clear and efficient business model (increasingly based on a data-driven economy) but are not able to deduce from this business model a clear digital governance model for the communities of network users that continuously feed their own business model;[21] (3) individuals, who are end-users of media companies and members of online communities, increasingly raise regulatory expectations that are addressed both to public and private actors:[22] however, fragmentation and inconsistency of such expectations prevents a coherent digital governance model from emerging out of online participation.

Following what we have remarked so far, we may observe that, at present, a new digital governance model needs to find out answers and regulatory solutions at two different but interconnected levels: (1) at the epistemological level, with regard to the proliferation of different epistemic models (artificial versus human intelligence; symbolics versus statistics; semantics versus syntax; optimization versus explanation; data-driven versus value-laden, etc.); (2) at the political level, with regard to the plurality of relevant political, economic, and social actors (e.g., national governments, supranational bodies, big tech, platform owners, online community of users, and citizens). It is becoming clear that we cannot separate the epistemological from the political level: the rise of computational power has connected them together in a much deeper, however less visible, way. As digital ICT has increasingly become embedded in society, the use of computational power continues to reshape our political, economic, and social environment.

As we have remarked, computational power affects our everyday forms of life. It is exactly in this regard that Wittgenstein's idea of a form of life becomes crucial, since it allows us to appreciate how hard it is to criticize what works successfully and is embodied in our forms of life: "As I read him, Wittgenstein thought that once a set of practices is ingrained enough to become your form of life, it is difficult to substantively criticise them or even to recognise them as what they are. That's because our form of life is 'what has to be accepted, the given.' We can no longer get outside of it."[23] What conditions and shapes our whole daily practices – that is, the set of syntactic and semantic rules that are conveyed and implemented throughout these practices – conditions and shapes our lives. This form of regulation – that is different from the traditional one but not less effective – operates neither from above nor from below but, so to speak, from within.

In this perspective, computational power aspires to be the new inner model of governance of our everyday forms of life: it does not aim to regulate society directly, formally, and vertically, but it aims to run how we take decisions, we anticipate and represent reality, we adapt to the world and to the representation of us produced and diffused by computational models. Computational power does not tell us what to do but how to live: it embodies norms into agents, products, practices, and systems that structure our everyday forms of life:

> Much will depend not only on how these technological innovations are operationalized from a scientific perspective over the coming years, but crucially also the grounding or embedding of such scientific advances in large-scale social systems. Advances in AI, machine learning and robotics are not simply the terrain of discovery, but fundamentally the ambivalent realm of lived experience, human experimentation and alternative social futures.[24]

This is the normative dimension of artificial agency, i.e., the fact of conveying (legal, moral, political, social, etc.) norms that are gradually removed from discussion, deliberation, interpretation, and evaluation. This reflective sphere of discussion, deliberation, interpretation, and evaluation is and remains crucial because, without it, there is no public, open, and shared mediation between means and ends, between technological operations and social responses. We cannot simply replace the correctness and fairness of a social response with the success and efficiency of a technological operation. As human beings, we have an inalienable role, in which our own singularity is revealed: to keep debating the meaning to be given to our everyday forms of life.

NOTES

1 Johann Wolfgang Goethe, *Faust* (New York: Anchor, 1990), 153.
2 On this Friedrich Nietzsche, *The Will to Power* (London: Penguin Classics, 2017); and Carl Schmitt, *The Tyranny of Values* (Washington: Plutarch Press, 1996).
3 Luciano Floridi et al., "AI 4 People – An Ethical Framework for a Good AI Society: Opportunities, Risks, Principles, and Recommendation," *Minds and Machines* 28, no. 4 (2018): 689 (we underline).
4 In this sense see Massimo Durante, *Computational Power. The Impact of ICT on Law, Society, and Knowledge* (London-New York: Routledge, 2021), Chapter 3.

5 On this point see Philipp Brey, "Values in Technology and Disclosive Computer Ethics," in *Information and Computer Ethics*, ed. Luciano Floridi (Cambridge: Cambridge University Press, 2010), 41–58.

6 In this sense see *David Weinberger, Everyday Chaos. Technology, Complexity, and How We're Thriving in a New World of Possibility* (Harvard: Harvard Business School, 2019), 62: "This reveals a weakness in our traditional basic strategy for managing what will happen, for the elements of a machine learning model may not have the sort of one-to-one relationship that we envision when we search for the right 'lever' to pull. When everything affects everything else, and when some of those relationships are complex and non-linear – that is, tiny changes can dramatically change the course of events – butterflies can be as important as levers. Overall, these changes mean that while models have been the stable frameworks that enable explanation, now we often explain something by trying to figure out the model our machines have created."

7 On this see, among others, Ugo Pagallo, "On the Principle of Privacy by Design and its Limits: Technology, Ethics and the Rule of Law," in *European Data Protection: In Good Health?* ed. Serge Gutwirth, Ronald Leenes, Paul De Hert, and Yves Poullet (Dordrecht: Springer, 2012), 331–346; Ann Cavoukian, "Privacy by Design: The Definitive Workshop," *Identity in Information Society* 3, no. 2 (2010): 247–251; Mireille Hildebrandt, "Legal Protection by Design: Objections and Refutations," *Legisprudence* 5 (2011): 223–248; Karen Yeung, "Towards an Understanding of Regulation by Design," in *Regulating Technologies: Legal Futures, Regulatory Frames and Technological Fixes*, ed. Roger Brownsword and Karen Yeung (London: Hart, 2007), 79–108; Michael Hunter Schwartz, "Teaching Law by Design: How Learning Theory and Instructional Design Can Inform and Reform Law Teaching," San Diego Law Review 38 (2001): 347–451.

8 Mireille Hildebrandt, "Legal and Technological Normativity: More (and Less) Than Twin Sisters," *Techné: Journal of the Society for Philosophy and Technology* 12, no. 3 (2008): 169–183; Mireille Hildebrandt, *Smart Technologies and the End(s) of Law: Novel Entanglement of Law and Technology* (Cheltenham: Edgar Elgar, 2016).

9 On this Luciano Floridi, *The Fourth Revolution: How the Infosphere is Reshaping the Human Reality* (Oxford: OUP, 2014).

10 In this perspective, Mireille Hildebrandt has recently claimed, in her book *Smart Technologies and the End(s) of Law* (Cheltenham: Edgar Elgar, 2016), that we are in transit from an information society to a data-driven society, which has far-reaching consequences for the world we depend on. Notably, she remarks how the pervasive employment of machine-learning technologies that inform so-called data-driven agency has a deep impact on law, since it threatens privacy, identity, autonomy, non-discrimination, due process, and the presumption of innocence.

11 Luciano Floridi, *The Fourth Revolution: How the Infosphere is Reshaping the Human Reality* (Oxford: OUP, 2014), 8: "[...] increasingly more power is available at decreasing costs, to ever more people, in quantities and at a pace that are mindboggling. The limits of computing power seem to be mainly physical. They concern how well our ICTs can dissipate heat and recover from unavoidable hardware faults while becoming increasingly small. This is the rocket that has made humanity travel from history to hyperhistory, to use a previous analogy." In this sense, see also Massimo Durante, "AI and Worldviews in the Age of Computational Power," in *The Routledge Social Science Handbook of AI*, ed. Anthony Elliott (London-New York: Routledge, 2021), Chapter 16.

12 On this see Massimo Durante, "AI and Worldviews in the Age of Computational Power," in *The Routledge Social Science Handbook of AI*, ed. Anthony Elliott (London-New York: Routledge, 2021), Chapter 1.

13 Massimo Durante, "AI and Worldviews in the Age of Computational Power," in *The Routledge Social Science Handbook of AI*, ed. Anthony Elliott (London-New York: Routledge, 2021), In

this sense, see also Anthony Elliott, *The Culture of AI: Everyday Life and the Digital Revolution* (Abingdon-New York: Routledge, 2019); and David Weinberger, *Everyday Chaos. Technology, Complexity, and How We're Thriving in a New World of Possibility* (Harvard: Harvard Business School, 2019), 62.

14 The point that tools and technologies are not something that are mere instrumental improvements but something that change the human condition can be already found in Martin Heidegger, *The Question Concerning Technology* (Harper, New York, 1977).

15 Michael Lynch, *The Internet of Us: Knowing More and Understanding Less in the Age of Big Data* (London-New York: Liveright Publisher, 2016), 10 (emphasis added).

16 I have analyzed these side effects, for instance, in relation to the governance model of digital memory and oblivion that stemmed from the *Google Spain* judgement of the Court of Justice of the European Union (C-131/12). On this see Massimo Durante, *Computational Power. The Impact of ICT on Law, Society, and Knowledge* (London-New York: Routledge, 2021), Chapter 4.

17 Jack Balkin, "Free Speech in the Algorithmic Society: Big Data, Private Governance, and New School Speech Regulation," *University of California, Davis Law Review* 51 (2018): 1149–1210.

18 We could further differentiate the internal composition of the three main actors involved in the digital governance as well as we could figure out the presence of some other actors in relation to the sector at stake (consider, for instance, the current complexity of the global and multilevel Internet governance with regard to the institutional role of ICANN and the multi-stakeholder society). On this issue, see Roxana Radu, *Negotiating Internet Governance* (Oxford: Oxford University Press, 2019); Milton Mueller, *Will the Internet Fragment? Sovereignty, Globalization and Cyberspace* (New York: Polity Press, 2017); Lee Bygrave, *Internet Governance by Contract* (Oxford: Oxford University Press, 2015).

19 Michel Serres, *Times of Crisis. What the Financial Crisis Revealed and How to Reinvest Our Lives and Future* (New York: Bloomsbury, 2015), 39.

20 We have the tendency to reduce these triadic or pluralistic relationships to the dualistic forms of top-down regulation (between public actors and individuals); bottom-up regulation (between individuals and private actors); or co-regulation (between public and private actors or between private actors and individuals). On the need for mediation between different forms of regulation see recently, Ugo Pagallo, Pompeu Casanovas, and Madelin Robert, "The Middle-out Approach: Assessing Models of Legal Governance in Data Protection, Artificial Intelligence, and the Web of Data," *The Theory and Practice of Legislation* 7, no. 1 (2019): 1–25.

21 Jack Balkin, "Free Speech in the Algorithmic Society: Big Data, Private Governance, and New School Speech Regulation," *University of California, Davis Law Review* 51 (2018): 1192: "Twenty-first century governors of digital speech, by contrast, make their money by facilitating and encouraging the production of content by ordinary people and governing the communities of speakers that result. New media companies like Facebook, Google, YouTube, and Twitter do not produce most of the content they serve. Rather, their business model requires them to induce as many people as possible around the world to post, speak, and broadcast to each other. Constant production of content by end-users, in turn, captures audience attention. This allows digital media companies to sell advertising, collect data about end-users, and use this information to sell even more advertising."

22 In this sense, Jack Balkin, "Free Speech in the Algorithmic Society: Big Data, Private Governance, and New School Speech Regulation," *University of California, Davis Law Review* 51 (2018): 1198: "Put another way, as online speech platforms govern, and increasingly resemble governments, it is hardly surprising that end-users expect them to abide by the basic obligations of those who govern populations in democratic societies."

23 Michael Lynch, *The Internet of Us: Knowing More and Understanding Less in the Age of Big Data* (London-New York: Liveright Publisher, 2016), 10, who cites Stanley Cavell, *Must We Mean What We Say?* (New York: Scribner and Sons, 1969), and Ludwig Wittgenstein, *Philosophical Investigations*, trans. G.E.M. Anscombe (Oxford: Basil Blackwell, 1963), 226.

24 Anthony Elliott, *The Culture of AI: Everyday Life and the Digital Revolution* (Abingdon-New York: Routledge, 2019), 201.

REFERENCES

Balkin, Jack, "Free Speech in the Algorithmic Society: Big Data, Private Governance, and New School Speech Regulation," *University of California, Davis Law Review* 51 (2018): 1149–1210.

Brey, Philipp, "Values in Technology and Disclosive Computer Ethics," in *Information and Computer Ethics*, ed. Luciano Floridi (Cambridge: Cambridge University Press, 2010), 41–58.

Bygrave, Lee, *Internet Governance by Contract* (Oxford: Oxford University Press, 2015).

Cavell, Stanley, *Must We Mean What We Say?* (New York: Scribner and Sons, 1969).

Cavoukian, Ann, "Privacy by Design: The Definitive Workshop," *Identity in Information Society* 3, no. 2 (2010): 247–251.

Durante, Massimo, "AI and Worldviews in the Age of Computational Power," in *The Routledge Social Science Handbook of AI*, ed. Anthony Elliott (London-New York: Routledge, 2021), Chapter 16.

Durante, Massimo, *Computational Power. The Impact of ICT on Law, Society, and Knowledge* (London-New York: Routledge, 2021).

Elliott, Anthony, *The Culture of AI. Everyday Life and the Digital Revolution* (Abingdon-New York: Routledge, 2019).

Floridi, Luciano, *The Fourth Revolution: How the Infosphere is Reshaping the Human Reality* (Oxford: OUP, 2014).

Floridi, Luciano, Josh Cowls, Monica Beltrametti, Raja Chatila, Patrice Chazerand, Virginia Dignum, Christoph Luetge, Robert Madelin, Ugo Pagallo, Francesca Rossi, Burkhard Schafer, Peggy Valcke, and Effy Vayena, "AI 4 People – An Ethical Framework for a Good AI Society: Opportunities, Risks, Principles, and Recommendation," *Minds and Machines* 28, no. 4 (2018): 689–707.

Goethe, Johann Wolfgang, *Faust* (New York: Anchor, 1990).

Heidegger, Martin, *The Question Concerning Technology* (New York: Harper, 1977).

Hildebrandt, Mireille, "Legal and Technological Normativity: More (and Less) Than Twin Sisters," *Techné: Journal of the Society for Philosophy and Technology* 12, no. 3 (2008): 169–183.

Hildebrandt, Mireille, "Legal Protection by Design: Objections and Refutations," *Legisprudence* 5 (2011): 223–248.

Hildebrandt, Mireille, *Smart Technologies and the End(s) of Law: Novel Entanglement of Law and Technology* (Cheltenham: Edgar Elgar, 2016).

Lynch, Michael, *The Internet of Us: Knowing More and Understanding Less in the Age of Big Data* (London-New York: Liveright Publisher, 2016).

Mueller, Milton, *Will the Internet Fragment? Sovereignty, Globalization and Cyberspace* (New York: Polity Press, 2017).

Nietzsche, Friedrich, *The Will to Power* (London: Penguin Classics, 2017).

Pagallo, Ugo, "On the Principle of Privacy by Design and its Limits: Technology, Ethics and the Rule of Law," in *European Data Protection: In Good Health?* ed. Serge Gutwirth, Ronald Leenes, Paul De Hert, and Yves Poullet (Dordrecht: Springer, 2012), 331–346.

Pagallo, Ugo, Pompeu Casanovas, and Robert Madelin, "The Middle-out Approach: Assessing Models of Legal Governance in Data Protection, Artificial Intelligence, and the Web of Data," *The Theory and Practice of Legislation* 7, no. 1 (2019): 1–25.

Radu, Roxana, *Negotiating Internet Governance* (Oxford: Oxford University Press, 2019).

Schmitt, Carl, *The Tyranny of Values* (Washington: Plutarch Press, 1996).

Schwartz, Michael Hunter, "Teaching Law by Design: How Learning Theory and Instructional Design Can Inform and Reform Law Teaching," *San Diego Law Review* 38 (2001): 347–451.

Serres, Michel, *Times of Crisis: What the Financial Crisis Revealed and How to Reinvest Our Lives and Future* (New York: Bloomsbury, 2015).

Weinberger, David, *Everyday Chaos. Technology, Complexity, and How We're Thriving in a New World of Possibility* (Harvard: Harvard Business School, 2019).

Wittgenstein, Ludwig, *Philosophical Investigations*, trans. G. E. M. Anscombe (Oxford: Basil Blackwell, 1963).

Yeung, Karen, "Towards an Understanding of Regulation by Design," in *Regulating Technologies: Legal Futures, Regulatory Frames and Technological Fixes*, ed. Roger Brownsword and Karen Yeung (London: Hart, 2007), 79–108.

CHAPTER 10

Law, Governance, and Artificial Intelligence – the Case of Intelligent Online Dispute Resolution

John Zeleznikow

AN INTRODUCTION TO ONLINE DISPUTE RESOLUTION

What is Online Dispute Resolution

Lodder and Zeleznikow indicate that while there is no generally accepted definition of Online Dispute Resolution (ODR), *we can think of it as using the Internet to perform Alternative Dispute Resolution (ADR).*[1] While this is a helpful working definition, it is important to note that one difficulty in providing a more precise and widely accepted definition is that ODR is many things, to many people.

ODR is often described as

1. Technology Assisted Dispute Resolution; or

2. Technology Facilitated Dispute Resolution; or

3. Technology Based Dispute Resolution

In dispute resolution, there are generally three parties, the participants in the conflict and an independent mediator. In ODR, there is a further common fourth party – the information technology component. ODR is a natural evolution of the trend toward using

DOI: 10.1201/9781003226406-11

alternative approaches to litigation across a wide range of civil, commercial, family, and other contexts.

ODR provides solutions for cases that do not justify long, complex trials. The unsatisfied purchaser of a Taylor Swift CD over the Internet is more likely to prefer an online process for achieving redress rather than pursuing litigation with the seller, who may be based in another country.

The earliest use of information technology in dispute resolution is the telephone.[2] It is a simple measure for convening people who cannot or should not be together in the same room, whether owing to geographical situations or to extremely vitriolic situations, or those where violence has occurred.

As Internet technology has become widespread, much attention has been directed at using these tools for dispute resolution. In some ways, ODR is a natural evolution of convening over the telephone. Information Technology offers parties different levels of immediacy, interactivity, and media richness to choose from. Through some platforms, parties can choose to communicate through text; through others, they can convene in real-time video, allowing them to see each other and, possibly, a mediator.

However, ODR is far more than a range of new communication platforms. When discussing ODR, one might be discussing any of the following: the online communication platform used for exchanging offers and messages in an ODR process; the individual processes from the ADR spectrum (e.g., online negotiation, online mediation, online early neutral evaluation, online arbitration); the ODR system – an environment in which parties to specific types of disputes are led through a particular process or set of processes on their way to a resolution. As opposed to an individual process, the system is a component of a larger environment.

ODR technology/software aims far beyond the "communications platforms" discussed above. ODR developers are seeking to create intelligent agents, and robust negotiation support systems (NSS). These systems aim to assist humans in achieving better outcomes than they would themselves, even when performing to the peak of their abilities.

Zeleznikow argues that the development of legal decision support systems has led to: consistency – by replicating the manner in which decisions are made, decision support systems are encouraging the spread of consistency in legal decision making; transparency – by demonstrating how legal decisions are made, legal decision support systems are leading to better community understanding of legal domains.[3] This has the desired benefit of decreasing the level of public criticism of judicial decision making; efficiency – one of the major benefits of decision support systems is to make firms more efficient; Enhanced support for dispute resolution – Users of legal decision support systems are aware of the likely outcome of litigation and thus are encouraged to avoid the costs and emotional stress of legal proceedings.[4] Such systems can also support users to engage in trade-offs.[5]

The Vanishing Trial

Galanter states that, in the United States, an abundance of data shows that the number of trials – federal and state, civil, and criminal, jury and bench – is declining.[6] The shrinking number of trials is particularly striking because virtually everything else in the legal world is growing – the population of lawyers, the number of cases, expenditures on law, the amount of regulation, the volume of authoritative legal material, and not least, the place of law, lawyers, and courts in public consciousness.

He provides empirical evidence that the portion of federal civil cases resolved by trial fell from 11.5% in 1962 to 1.8% in 2002, continuing a long historic decline.[7] More startling was the 60% decline in the absolute number of trials since the mid-1980s. The makeup of trials shifted from a predominance of torts to a predominance of civil rights, but trials are declining in every case category. A similar decline in both the percentage and the absolute number of trials is found in federal criminal cases and in bankruptcy cases. The phenomenon is not confined to the federal courts; there are comparable declines of trials, both civil and criminal, in the state courts, where the great majority of trials occur. While plausible causes for this decline include a shift in ideology and practice among litigants, lawyers, and judges, another significant manifestation of this shift is the diversion of cases to Alternative Dispute Resolution (ADR) forums. Within the courts, judges conduct trials at only a fraction of the rate that their predecessors did, but they are more heavily involved in the early stages of cases. Although virtually every other indicator of legal activity is rising, trials are declining not only in relation to cases in the courts but also to the size of the population and the size of the economy.

The Australian High Court's view on Artificial Intelligence in Law

In an address to the Australian Disputes Centre annual ADR Awards Night in Sydney Australia on August 10, 2017, the then Chief Justice of the Australian High Court, Hon. Robert French AM, said:[8]

> Nowhere is the potential for change more dramatic than in the use of technology and, in particular, artificial intelligence. An immediate application is online dispute resolution' using what has been described as 'a virtual space in which disputants have a variety of dispute resolution tools at their disposal.

He cited the three-step model of Lodder and Zeleznikow.[9] Justice French states further that

> There are challenges in connection with the use of artificial intelligence in this area, especially in relation to machine-based application of legal rules whether they be statutory or common law. Such rules are rarely unambiguous and generally offer constructional choices which don't readily translate into machine logic. An alternative approach is called a 'Data-centric approach'. The relevant computer is provided with data about the facts and outcomes of a large number of cases on the basis of which it is asked to estimate the probabilities of outcomes given a particular set of facts and relevant legal issues. Such a tool might be useful as a kind of surrogate early neutral evaluator.

He concludes by arguing

> There are, of course, issues of justice which transcend the negotiating objectives of the parties, particularly in family law disputes where the law requires that the interests of affected children be treated as paramount. And in complex

multi-party negotiations such as environmental or native title disputes, the question of the public interest is a very large aspect of the context in which negotiations must be undertaken. The challenge is to take the benefits of technology without compromising the essential characteristics of courts in terms of independence, openness, fairness and accountability through the provision of reasoned decisions.

THE EVOLUTION OF ONLINE DISPUTE RESOLUTION

Advances in Dispute Resolution and Information Technology over the past 50 years have led to the current evolution of ODR. The decade of the 1970s saw the rise of the ADR movement. Modern alternatives to litigation were heavily influenced by the National Conference on the Causes of Popular Dissatisfaction with the Administration of Justice, which took place in Minneapolis, Minnesota from April 7 to 9 1976. At this conference, then USA Chief Justice Warren Burger, encouraged the exploration and use of informal dispute resolution processes.[10] Sander introduced the idea of the Multidoor Courthouse movement.[11] A few years later we had the publication of Fisher and Ury's (1981) *Getting to Yes*[12] book and Howard Raiffa's 1982 book *The Art and Science of Negotiation*,[13] both arising from the Harvard Program on Negotiation.[14]

The decade of the 1980s saw the development (and hype about) expert systems to model legalistic decision making. It was proposed that eventually such expert systems could change the nature of legal practice. An example of such systems includes TAXMAN[15] and the Latent Damage Advisor of Capper and Susskind.[16]

The decade of the 1990s saw the development of the Internet and initial proposals for ODR. Much of this work came from legal academics rather than technology developers. They saw the potential of ODR to resolve disputes that originated on the Internet.[17]

The twenty-first century saw the development of ODR for E-commerce; in particular its use by EBay and PayPal.[18] The past decade has seen the development of practical usable systems such as Rechtwijzer in the Netherlands[19] and the Civil Resolution Tribunal in British Columbia.[20] These systems will be discussed are discussed later in the paper.

Recently, ODR has moved beyond E-commerce. ODR is finally being used for non-financial disputes – see for instance the book of Katsh and Rabinovich-Einy[21] and the access to justice work at Kent Law School.[22]

In the next section we discuss examples of two systems developed to support decision making in Australian Family Law. These systems were constructed and utilized more than 15 years ago. Susskind discusses two tiers of online courts.[23] In the second tier he imagines a machine learning system helping parties by predicting the likely outcome of their case were it to come before a human judge. There is no need to hypothesize such an outcome – it occurred in 1995 in the Split-Up system of Zeleznikow and Stranieri.[24]

ARTIFICIAL INTELLIGENCE TO SUPPORT ONLINE FAMILY DISPUTE RESOLUTION

Zeleznikow notes that: "Despite the concept of negotiation having a long and varied history, there is limited theory on what are good principles to use when conducting a negotiation.

And the theories that do exist have often been developed for specific domains."[25] Such theories were first developed in the 1960s, at the same time as the first development of Artificial Intelligence systems such as the Eliza program of Weizenbaum.[26] Eliza was a rule-based expert system.

In the rule-based approach,[27] the knowledge of a specific legal domain is represented as a collection of rules of the form

IF <condition(s)> THEN action/conclusion.

For example, consider the domain of driving offenses, in Victoria, Australia. Drivers can lose their license either by being drunk while driving or exceeding a specified number of points in a given time. More specifically, probationary drivers (those who have held a driver's license for less than three years) are not permitted to have even a trace of any alcohol in their blood. Other drivers must have a blood alcohol level not exceeding 0.05%. This knowledge can be modeled by the following rules:

- (a) IF drive(X) & (blood_alcohol(X) > .05) & (license(X) >= 36) THEN licence_loss(X);
- (b) IF drive(X) & (blood_alcohol(X) > .00) & (license(X) < 36) THEN licence_loss (x).

Case-based reasoning is the process of using previous experience to analyze or solve a new problem, explain why previous experiences are or are not similar to the present problem and adapting past solutions to meet the requirements. Precedents play a more central role in Common law than in Civil law. Using the principle of *stare decisis*, to decide a new case, legal decision-makers search for the most similar case decided at the same or higher level in the hierarchy.

Stranieri and Zeleznikow showed that machine learning could be gainfully used to model legal reasoning. Machine learning is that subsection of learning in which the artificial intelligence system attempts to learn automatically.[28]

In the Split-Up system[29] (Stranieri et al. 1999) provided advice about the distribution of marital property following divorce in Australia by using machine learning to offer advice about BATNAs[30] (a BATNA is used to inform disputants of the likely outcome if the dispute were it to be decided by decision-maker e.g., judge, arbitrator, or ombudsman) re the distribution of marital property following divorce (Figure 10.1).

Despite using Artificial Intelligence, it involved much conceptual modeling. Ninety-four Toulmin argument structures were developed to model the domain as it existed in 1995.[31] A subset is shown in Figure 10.1. Twenty-five years later, the theoretical principles behind artificial intelligence software have not changed. But computer software and hardware are much less expensive, and data can be much more easily stored.

The Split-Up was the focus of media discussion when it was used to advise upon the divorce of Prince Charles and Lady Diana of the United Kingdom. It featured in the London Daily Telegraph of July 4, 1996, and in a ten-minute program on GTV9's (Australia) Current Affair on Monday 26 August 1996.[32]

Initially the developers were reluctant to provide advice upon the marriage of Prince Charles and Lady Diana because the system provides advice about Australian Family Law.

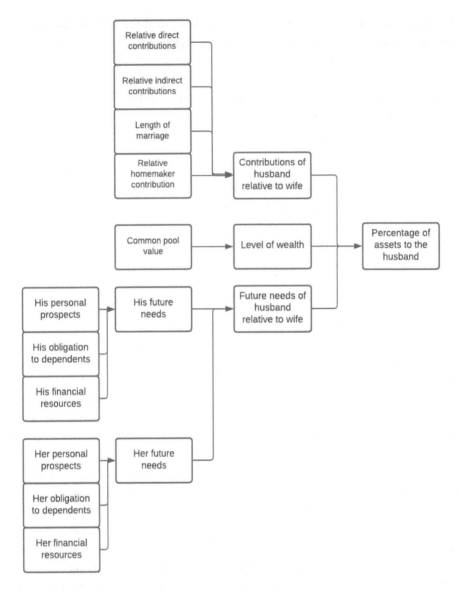

FIGURE 10.1   The Toulmin Argument structure in split-up.

Charles and Diana lived in England; the developers of the software did not know about the assets of the couple; one of the major reasons that the royal couple wanted to avoid litigation was to keep their assets secret; and the Split-Up system performs machine learning on commonplace or ordinary cases. The royal divorce did not meet the definition of a commonplace case.[33]

When the system was used, Lady Diana was classified as a single mother who had lost her job. As such, even with the prospect of a plentiful distribution of property toward her, the system awarded her 70% of the common pool. A useful article on the general use of software can be found in an article in the Economist of March 12, 2005 with the title "AI am the Law."[34]

Portable and the Legal Services Commission of South Australia designed and developed Amica,[35] a digital solution for Australian separating couples. Amica includes a machine learning algorithm that provides a suggested division of a former couple's total assets. In this capacity, AMICA emulates the Split-Up.

Although the Split-Up system provides advice about BATNAs, it is not a Negotiation Support System. Bellucci and Zeleznikow[36] used game theory[37] as developed by John Nash[38] to provide advice to disputing parents on how they could best negotiate trade-offs. The disputing parties were asked to indicate how much they valued each item in dispute. Using logrolling,[39] parties obtained what they desired most.

The Family Winner software won its heat of the ABC (Australia) TV series science show The New Inventors.[40] It was also heavily featured in the Economist article on the March of the Robolawyers,[41] published on March 12, 2006. The article says

> Family Winner, guides the couple through a series of trade-offs and compensation strategies. According to John Zeleznikow, a computer scientist at Victoria University in Melbourne, who developed the software with his colleague Emilia Bellucci, the results are desirable because both parties end up with what they value most.

The Family Winner software was tested upon 50 divorcing couples, with the outcomes evaluated by Victoria Legal Aid. Each party was given a limited number of points, which they are asked to allocate to the items of property they wish to keep. Through a multi-step process of modification, the parties are encouraged to give priority to the items they most value. It was found that, using the software, each party ended up with 70–80% of what they originally wanted, rather than the usual 50–50 split.

Richard Susskind, who was and still is technology adviser to England's Lord Chief Justice claimed,

> In the eyes of the law, this approach to conflict resolution would appear to be superior because it produces a more equitable outcome than the courts. But, it can only be employed only if both parties consent to its use, which is unlikely in the most acrimonious cases. Sadly, divorcing couples do not necessarily always want what is fair. The system is intended for use by couples seeking out-of-court mediation as they separate, rather than those already involved in legal battle.

SOME ONLINE DISPUTE RESOLUTION SYSTEMS WHICH MOVE BEYOND E-COMMERCE

Given that ODR has moved beyond E-commerce and is finally being used for non-financial disputes, let us first investigate some non-E-commerce based ODR systems.

Non-E-commerce-Based ODR Systems

Our Family Wizard is an electronic posting service and a tool that can provide verifiable evidence of how parental communication takes place. It assists separating parents to engage in appropriate and civil behavior while also helping develop parenting planning and maintaining a record of parent behavior. Such record keeping, which can be viewed by judges and other court officers, encourages appropriate behavior. While Our Family

Wizard is a useful tool, it provides limited support to help disputants resolve their conflicts on par with other current ODR systems.[42]

Adieu Technologies offers family law advice and supports triaging and drafting plans.[43] One of its agents, Lumi, is a bot with expertise in law, mediation, and counseling. After having a confidential conversation with a client, Lumi will create a step-by-step plan to help the client navigate the mediation.

The Hague Institute for International Law which developed the Rechtwjjzer system says: "Platforms for online dispute resolution are still scarce. Ideally, they integrate the three main stages of problem-solving: self-help, facilitation by a bridge-builder and coming to decisions under supervision of a judge."[44] Rechtwijzer offers advice about Dutch Family Law. It provides support for entering data and forms, and advice about procedures and outcomes. Sadly, the initial project proved financially unsustainable. The three-year collaboration between the Hague Institute for the Internationalization of Law (HIIL), the Dutch Legal Aid Board and the US firm Modria was dissolved in 2017. But as with the failure to commercialize both Split-Up and Family Winner, this failure had little to do with the quality of the systems. Rather, the systems were abandoned because of financial maneuvers. In the case of Rechtwijzer, Modria was bought by the US Legal Technology firm Tyler.

The British Columbia Civil Resolution Tribunal,[45] comes closest to providing a full suite of dispute resolution services. The process commences with Solution Explorer. It diagnoses the dispute and provides legal information and tools such as customized letter templates. The template is essentially a formal, legal looking, letter of demand. If this action does not resolve the dispute, one can then apply to the Civil Resolution Tribunal for dispute resolution.

The system directs the user to the appropriate application forms. Once the application is accepted, the user enters a secure and confidential negotiation platform, where the disputants can attempt to (by themselves) resolve their dispute. If the parties cannot resolve the dispute, a facilitator will assist. Agreements can be turned into enforceable orders. If negotiation or facilitation does not lead to a resolution, an independent member will make a determination about the dispute.

The Civil Resolution Tribunal deals with the following categories of cases:

1. **Motor vehicle injury** disputes up to C$50,000

2. **Small claims** disputes up to C$5,000

3. **Strata property** (condominium) disputes of any amount

4. **Societies and cooperative associations** disputes of any amount

5. **Shared accommodation and some housing** disputes up to C$5,000

For these five domains, potential litigants in British Columbia can only use the Civil Resolution Tribunal. No paper-based solutions are available. Digitally disadvantaged litigants are provided with assistance in accessing the Internet.

One of the major reasons that the Civil Resolution Tribunal has been so successful, is that British Columbia residents are mandated to use the system when dealing with issues listed. above. Because most litigants presenting cases in small claims tribunals are self-represented,[46] the development of advisory systems for self-represented litigants is vital. But how can this be achieved?

Software such as Family Winner and Smartsettle[47] assist parties with trade-offs. Lodder and Zeleznikow claim the ideal ODR system should be a hybrid system offering case management, BATNA advice, communication support and tools that advise upon trade-offs. No current systems provide such hybrid support. Schmitz and Zeleznikow discuss this in detail.[48]

Classifying ODR Systems

Thiessen and Zeleznikow[49] observed ODR systems could be classified into seven categories as indicated in Table 10.1. The systems they examined represented a wide range of approaches to dispute resolution (e.g., Artificial Intelligence, Social Psychology, and Game Theory). Jennings et al. noted that given the wide variety of possibilities for building such systems, there is no universally best approach or technique.[50] Rather, there is an eclectic bag of methods with properties and performance characteristics that vary significantly depending on the context.

What all of the selected ODR Systems have in common is that they provide an alternative to litigation providing a mechanism by which parties involved in a dispute can communicate over the Internet. Many of the illustrated systems were specifically designed to provide the best approach for a particular path to resolution.

The various categories of these systems, the techniques that these systems use and examples of software in each of these categories are indicated in Table 10.1.

TABLE 10.1 Categorization of ODR Systems

Category	Methods	Main Players
Information Systems	Provision of information that parties can use to resolve their dispute	Provision of information that parties can use to resolve their dispute
Univariate blind bidding	Automation for single monetary issues	CyberSettle
Document management for negotiation	Facilitators working online and/or offline with parties making use of formal structured document management tools to help them create their contract	Negoisst
eNegotiation (or automated mediation) systems	Sophisticated optimization algorithms to generate optimal solutions for complex problems	FamilyWinner Inspire SmartSettle
Customized for negotiation or mediation of a particular type of dispute	Automated negotiation with structured forms	eBay UPI SquareTrade
General virtual mediation rooms	Human mediators working online with parties making use of mediums such as email, instant messengers, telephone, and discussion forums	ECODIR Mediation Room SquareTrade
Arbitration Systems	Human arbitrators working online with parties making use of mediums such as email, instant messengers, telephone, and discussion forums	Word & Bond

The Thiessen and Zeleznikow model focus upon the design of Information Systems, rather on the needs of users. The three-step model of Lodder and Zeleznikow is designed for neither users nor information systems designers.[51] It integrates the research experience of the two developers of the theory.

Zeleznikow claims that a truly helpful ODR system should provide case management, triaging, advisory tools, communication tools, decision support tools, and drafting software.[52]

An ODR system should allow users to enter information, ask them for appropriate data and provide for templates to initiate the dispute. As an example, currently most clients of Victoria Legal Aid phone the organization to seek help. This process is expensive, time consuming, and often inaccurate for telephonists to enter data. Mistakes are often made. Potential litigants should be able to initiate the dispute, enter their pertinent data and track what is happening during the dispute as well as being aware of what documents are required at specific times. Because potential self-represented[53] litigants have a limited knowledge of legal processes, such support could be vital.

The ODR system should make triaging decisions, deciding how important it is to act in a timely manner and where to send the dispute. This may be particularly important in cases of domestic abuse or where there is a potential for children to be kidnapped. One would not expect self-represented litigants to use triaging systems. But triaging systems are vital for expediting action in high-risk cases. Such systems are thus vital for protecting the interests of at risk self-represented litigants.

An ODR system should provide tools for reality testing. Such tools could include, books, articles, reports of cases, copies of legislation, and videos; there would also be calculators (such as to advise upon child support) and BATNA advisory; systems (to inform disputants of the likely outcome if the dispute were to be decided by decision-maker e.g. judge, arbitrator, or ombudsman). Other useful advice that could be included are copies of the relevant Acts, links to landmark cases, relevant books and reports and videos providing useful parenting advice. Advisory tools (as suggested by Zeleznikow) are a vital cog.[54] An important associated question is how can we design advisory tools that self-represented litigants can gainfully use? Are the legal concepts behind these tools too difficult for amateurs to understand? How do we construct suitable user interfaces? This will be the subject of much future research.

Since the onset of the COVID-19 pandemic in early 2020, there has been a widespread use of communication tools in the provision of all forms of legal services as well as more widely.[55] ODR systems require communication tools for negotiation, mediation, conciliation, or facilitation. This could involve shuttle mediation if required. For many ODR providers, the provision of communication tools is their main goal. Wilson-Evered and Zeleznikow describe how Relationships Australia Queensland built a Family ODR system that used AdobeConnect to emulate Australian Online Family Dispute Resolution. Online communication tools will be important for potential litigants.[56]

If disputants cannot resolve their conflict, ODR software using game theory or artificial intelligence can be used to facilitate trade-offs. Family Winner and Smartsettle provide such services. Professionals (such as lawyers) can provide useful advice regarding trade-offs. In their absence, suitable decision support tools are vital.

Once an in-principle agreement has been reached by parties using an ODR system, software can be used to draft suitable agreements. Drafting plans (such as parenting plans) once there is an in-principle agreement for a resolution of a dispute, is a non-trivial task. And the task is of course more difficult when one or two of the litigants are not represented.

No single dispute is likely to require all six processes. The development of such a hybrid ODR system, however, would be very significant, but costly and very time and resource consuming. Such a system would be an important starting point for expanding into a world where Artificial Intelligence, however, is gainfully used.

THE GOVERNANCE OF INTELLIGENT ONLINE DISPUTE RESOLUTION SYSTEMS

Ebner and Zeleznikow examine the governance of ODR. They argue that the ODR environment is similar to the Wild West.[57] Unless there is appropriate governance of ODR, users may have very little trust in the processes and hence be reluctant to engage with them.

ODR generally knows little or no regulation, authority, standards, or monitoring. Within the field itself, governance is virtually non-existent: no body monitors quality control, no well-recognized and accepted standards of practice exist, no-one deals with complaints or investigates bad practices. Because ODR practice is by nature Internet-based, any external supervision, such as that stemming from consumer protection laws, is weak, and subject to jurisdictional shortcomings.

Most of the early ODR providers, and many of the current ones, have chosen the organizational/commercial model of *service provider company*. In this model, the management team behind the company is not personally identified (or even named at all). Many, perhaps most, service providers do not relate substantively to the identity, qualifications, or training of their neutrals. They do not share their code of practice, even if they have one; they do not share whether they subscribe to codes of practice formulated by organizations in other fields.

Ebner and Zeleznikow consider potential governance models for ODR systems.[58] A no-governance model views ODR as a free-market domain where users need to be aware of potential hazards. In a self-governance model, guidelines for practice are made voluntarily available and are therefore adhered to in a self-enforced manner with no external monitoring body present.

A third possibility would be an internal governance model run by industry groups. When such a model was proposed by Ebner and Zeleznikow in 2016, there were no suitable ODR Industry groups.[59] In Australia, there are professional bodies that could offer governance in IT (such as the Australian Computer Society) and ADR (the Resolution Institute). But there was no appropriate body for ODR. Standards would need to be implemented and enforced through monitoring and accreditation schemes. But as Wing et al. (2021) say

> There are an increasing number of efforts to create ODR guidance and standards by membership organizations, private enterprises, and government agencies, Some examples include a set of ODR Standards promulgated by the International Council for Online Dispute Resolution (2017) and a collaboration that is underway between the American Bar Association Dispute Resolution Section, the International Council

for Online Dispute Resolution, and the National Center for Technology and Dispute Resolution to develop robust standards and guidance on the application of technology to dispute resolution systems and software design, *platform management, and practitioner behavior.*[60]

An external governance model of ODR would require courts or governments to regulate ODR – an unlikely occurrence.

These models are proposed as a platform from which to initiate further discussion around the future of governance in ODR. To be accepted by users, ODR providers must subscribe to a rigorous governance model.

CONCLUSION

Artificial Intelligence has been gainfully used in health and engineering. Despite Artificial Intelligence and Law being first developed as a discipline almost 60 years ago, practical applications have been very slow in coming forward. One area that has become significant is the use of Artificial Intelligence in Online Dispute Resolution.

Until recently, Online Dispute Resolution has focused upon video-conferencing. It has tended to ignore the use of artificial intelligence including rule-based reasoning, case-based reasoning, game theory, and machine learning. Three classification systems for Online Dispute Resolution are discussed: Thiessen and Zeleznikow's classification with regard to the underlying information systems being used; Lodder and Zeleznikow's three-step model for using artificial intelligence and Zeleznikow's approach for developing user-centric and friendly systems. This third approach advocates using case management, triaging, providing advisory and decision support tools and drafting software as well as the constant feature of all current Online Dispute Resolution Systems – communication tools. Together with these frameworks we have discussed numerous current Online Dispute Resolution Tools – including the British Columbia Civil Resolution Tribunal and Rechtwijzer.

Using Artificial Intelligence can lead to a variety of governance and ethics issues. We concluded by investigating four potential governance models for Online Dispute Resolution.

NOTES

1 Arno R. Lodder, and John Zeleznikow, *Enhanced Dispute Resolution Through the Use of Information Technology* (Cambridge, New York: Cambridge University Press, 2010).

2 Mark Thomson, "Alternative Modes of Delivery for Family Dispute Resolution: The Telephone Dispute Resolution Service and the online FDR Project," *Journal of Family Studies* 17, no. 3 (2011): 253–257.

3 John Zeleznikow, "Using Web-based Legal Decision Support Systems to Improve Access to Justice," *Information and Communications Technology Law* 11, no. 1 (2002): 15–33.

4 Andrew Stranieri, John Zeleznikow, Mark Gawler, and Bryn Lewis, "A Hybrid Rule–neural Approach for the Automation of Legal Reasoning in the Discretionary Domain of Family Law in Australia," *Artificial Intelligence and Law* 7, no. 2–3 (1999): 153–183.

5 Emilla Bellucci, and John Zeleznikow, "Developing Negotiation Decision Support Systems that Support Mediators: A Case Study of the Family Winner System," *Journal of Artificial Intelligence and Law* 13, no. 2 (2006): 233–271.

6 Mark Galanter, "The Hundred-year Decline of Trials and the Thirty Years War," *Stanford Law Review* 57, no. 5 (2005): 1255–1274.

7   Marc Galanter, "The Vanishing Trial: An Examination of Trials and Related Matters in Federal and State Courts," *Journal of Empirical Legal Studies* 1, no. 3 (2004): 459–570.

8   See https://www.disputescentre.com.au/wp-content/uploads/2017/09/Robert-French-Address-Australian-Disputes-Centre-ADR-Award-Evening-10-8-.pdf, last viewed January 4, 2020.

9   Arno R. Lodder and John Zeleznikow, "Developing an Online Dispute Resolution Environment: Dialogue Tools and Negotiation Systems in a Three Step Model," *The Harvard Negotiation Law Review* 10 (2005): 287–338.

10 According to the United States Department of Justice, the term multi-door courthouse describes courts that offer an array of dispute resolution options or screen cases and then channel them to particular alternative dispute resolution processes.

11 Frank E. A. Sander, "The Multi-door Courthouse: Settling Disputes in the Year 2000," *Barrister* 3 (1976): 18–42.

12 Roger Fisher and William L. Ury, *Getting to Yes: Negotiating Agreement without Giving In* (New York: Penguin Group, 1981).

13 Howard Raiffa, *The Art and Science of Negotiation* (Cambridge, MA: Harvard University Press, 1982).

14 See https://www.pon.harvard.edu/, last viewed November 10, 2019.

15 L. Thorne McCarty, "Reflections on TAXMAN: An Experiment in Artificial Intelligence and Legal Reasoning," *Harvard Law Review* 90, no. 5 (1976): 837–893.

16 Richard E. Susskind, *Expert Systems in Law* (Oxford: Clarendon Press, 1987).

17 M. Ethan Katsh, "Dispute Resolution in Cyberspace," *Connecticut Law Review* 28 (1995): 953.

18 Colin Rule, *Online Dispute Resolution for Business: B2B, Ecommerce, Consumer, Employment, Insurance, and Other Commercial Conflicts* (San Francisco, CA: John Wiley & Sons, 2003); Colin Rule and Larry Friedberg, "The Appropriate Role of Dispute Resolution in Building Trust Online," *Artificial Intelligence and Law* 13, no. 2 (2005): 193–205.

19 Roger Smith, "Ministry of Justice for England and Wales Dives into the Deep Water on Online Dispute Resolution," *Dispute Resolution Magazine* 23 (2016): 28.

20 Shannon Salter and Darin Thompson, "Public-centered Civil Justice Redesign: A Case Study of the British Columbia Civil Resolution Tribunal," *McGill Journal of Dispute Resolution* 3 (2016): 113–136.

21 M. Ethan Katsh and Orna Rabinovich-Einy, *Digital Justice: Technology and the Internet of Disputes* (Oxford: Oxford University Press, 2017).

22 See https://www.kentlaw.iit.edu/institutes-centers/center-for-access-to-justice-and-technology, last viewed November 10, 2019.

23 Richard E. Susskind, *Online Courts and the Future of Justice* (Oxford: Oxford University Press, 2019).

24 John Zeleznikow and Andrew Stranieri, "The Split-up System: Integrating Neural Networks and Rule-based Reasoning in the Legal Domain," in *Proceedings of the 5th International Conference on Artificial Intelligence and Law* (New York: Association of Computing Machinery, 1995), 185–194.

25 John Zeleznikow, "The Challenges of Using Online Dispute Resolution to Support Self Represented Litigants," *Journal of Internet Law* 23, no. 7 (2020): 3–14.

26 Joseph Weizenbaum, "ELIZA – A Computer Program for the Study of Natural Language Communication between Man and Machine," *Communications of the ACM* 9, no. 1 (1966): 36–45.

27 For an excellent introduction to the use of Artificial Intelligence in Law, see John Zeleznikow and Dan Hunter, *Building Intelligent Legal Information Systems: Representation and Reasoning in Law (No. 13)* (Deventer, Bosten: Kluwer Law and Taxation Publishers, 1994).

28 Andrew Stranieri and John Zeleznikow, *Knowledge Discovery from Legal Databases (Vol. 69)* (Springer Science & Business Media, 2006).

29 Andrew Stranieri, John Zeleznikow, Mark Gawler, and Bryn Lewis, "A Hybrid Rule–neural Approach for the Automation of Legal Reasoning in the Discretionary Domain of Family Law in Australia," *Artificial Intelligence and Law* 7, no. 2–3 (1999): 153–183.

30 A BATNA is your best alternative to a negotiated agreement. The reason you negotiate with someone is to produce better results than would otherwise occur. If you are unaware of what results you could obtain if the negotiations are unsuccessful, you run the risk of entering into an agreement that you would be better off rejecting; or rejecting an agreement you would be better off entering into.

31 Stephen E. Toulmin, *The Uses of Argument* (Cambridge: Cambridge University Press, 1958).

32 The video can be seen at https://www.youtube.com/watch?v=u7A3H4lUjzM&t=0s, last viewed 5 August 2020.

33 Andrew Stranieri, John Zeleznikow, Mark Gawler, and Bryn Lewis, "A Hybrid Rule–neural Approach for the Automation of Legal Reasoning in the Discretionary Domain of Family Law in Australia," *Artificial Intelligence and Law* 7, no. 2–3 (1999): 153–183.

34 See https://www.economist.com/technology-quarterly/2005/03/12/ai-am-the-law, last accessed January 3, 2020.

35 See https://www.amica.gov.au/, last viewed 20 July 2020.

36 Emilla Bellucci, and John Zeleznikow, "Developing Negotiation Decision Support Systems that Support Mediators: A Case Study of the Family Winner System," *Journal of Artificial Intelligence and Law* 13, no. 2 (2006).

37 Game theory is a branch of applied mathematics that provides advice about the optimal distribution of resources. In the case of a negotiation, the goal of game theory is to develop the best outcome related to the choices each person has made.

38 Nash, who suffered from schizophrenia won the Nobel Prize for Economics in 1994. His ground-breaking work on game theory was published forty years earlier: John Nash, "Two-person cooperative games," *Econometrica: Journal of the Econometric Society* 21, no. 1 (1953): 128–140. A detailed examination of the use of artificial intelligence in negotiation can be found in Arno R. Lodder, and John Zeleznikow, *Enhanced Dispute Resolution Through the Use of Information Technology* (Cambridge, New York: Cambridge University Press, 2010).

39 Logrolling is a process in which participants look collectively at multiple issues to find issues that one party considers more important than does the opposing party. Logrolling is successful if the parties concede issues to which they give low importance values.

40 For a video of the show see https://www.youtube.com/watch?v=YOZczuvrou4&t=61s, last viewed November 10, 2019.

41 See https://www.economist.com/technology-quarterly/2006/03/11/march-of-the-robolawyers, last viewed January 3, 2020.

42 Hon. Thomas Trent Lewis, "Helping Families by Maintaining a Strong Well-Funded Family Court that Encourages Consensual Peacemaking: A Judicial Perspective," *Family Court Review* 53, no. 3 (2015): 371–377; Allen E. Barsky, "The Ethics of App-assisted Family Mediation," *Conflict Resolution Quarterly* 34, no. 1 (2016): 31–42.

43 See https://www.adieu.ai/, last viewed May 16, 2021.

44 See https://www.hiil.org/wp-content/uploads/2018/11/HiiL-Understanding-Justice-Needs-The-Elephant-in-the-Courtroom.pdf, last viewed November 10, 2019.

45 See https://civilresolutionbc.ca/, last viewed November 10, 2019.

46 Indeed, in many such tribunals/courts, legal representation is not permitted.

47 See https://smartsettle.com/ last viewed November 10, 2019.

48 Arno R. Lodder, and John Zeleznikow, *Enhanced Dispute Resolution Through the Use of Information Technology* (Cambridge, New York: Cambridge University Press, 2010); Amy J. Schmitz and John Zeleznikow, "Intelligent Legal Tech to Empower Self-Represented Litigants," *Columbia Science and Technology Law Review* 23, no. 1 (2022): 142–190.

49 Ernest Thiessen and John Zeleznikow, "Technical Aspects of Online Dispute Resolution Challenges and Opportunities," in *Proceedings of the Third Annual Forum on Online Dispute Resolution*, ed. Melissa Conley Tyler, Ethan Katsh, and Daewon Cho (Melbourne, 2004), 5–6.

50 Nicholas Robert Jennings, Peyman Faratin, Alessio R. Lomuscio, Simon Parsons, Carles Sierra, and Michael J. Wooldridge, "Automated Negotiation: Prospects, Methods and Challenges," *International Journal of Group Decision and Negotiation* 10, no. 2 (2001): 199–215.

51 Arno R. Lodder and John Zeleznikow, "Developing an Online Dispute Resolution Environment: Dialogue Tools and Negotiation Systems in a Three Step Model," *The Harvard Negotiation Law Review* 10 (2005): 287–338.

52 John Zeleznikow, "Using Artificial Intelligence to provide Intelligent Dispute Resolution Support," *Group Decision and Negotiation* (2021): 1–24.

53 A pro se or self-represented litigant is one who does not retain a lawyer and appears for himself in court. See Stephan Landsman, "The Growing Challenge of Pro se Litigation," *Lewis & Clark Law Review* 13, no. 2 (2009): 439–460.

54 John Zeleznikow, "Using Web-based Legal Decision Support Systems to Improve Access to Justice," *Information and Communications Technology Law* 11, no. 1 (2002): 15–33.

55 Tania Sourdin and John Zeleznikow, "Courts, Mediation and COVID-19," *Australian Business Law Review* 48, no. 2 (2020): 138–158.

56 Elisabeth Wilson-Evered and John Zeleznikow, *Online Family Dispute Resolution: Evidence for Creating the Ideal People and Technology Interface (Vol. 45)* (New York: Springer Nature, 2021).

57 Noam Ebner and John Zeleznikow, "No Sheriff in Town: Governance for Online Dispute Resolution," *Negotiation Journal* 32, no. 4 (2016): 297–323.

58 Noam Ebner and John Zeleznikow, "No Sheriff in Town: Governance for Online Dispute Resolution," *Negotiation Journal* 32, no. 4 (2016): 297–323.

59 Noam Ebner and John Zeleznikow, "No Sheriff in Town: Governance for Online Dispute Resolution," *Negotiation Journal* 32, no. 4 (2016): 297–323.

60 Leah Wing, Janet Martinez, Ethan Katsh, Colin Rule, "Designing Ethical Online Dispute Resolution Systems: The Rise of the Fourth Party," *Negotiation Journal* 37, no. 1 (2021): 49–64.

REFERENCES

Barsky, Allan E., "The Ethics of App-assisted Family Mediation," *Conflict Resolution Quarterly* 34, no. 1 (2016): 31–42.

Bellucci, Emilla, and John Zeleznikow, "Developing Negotiation Decision Support Systems that support mediators: A Case Study of the Family_Winner System," *Journal of Artificial Intelligence and Law* 13, no. 2 (2006): 233–271.

Ebner, Noam, and John Zeleznikow, "No Sheriff in Town: Governance for Online Dispute Resolution," *Negotiation Journal* 32, no. 4 (2016): 297–323.

Fisher, Roger, and William L. Ury, *Getting to Yes: Negotiating Agreement Without Giving In* (New York: Penguin Group, 1981).

Galanter, Marc, "The Vanishing Trial: An Examination of Trials and Related Matters in Federal and State Courts," *Journal of Empirical Legal Studies* 1, no. 3 (2004): 459–570.

Galanter, Mark, "The Hundred-year Decline of Trials and the Thirty Years War," *Stanford Law Review* 57, no. 5 (2005): 1255–1274.

Jennings, Nicholas Robert, Peyman Faratin, Alessio R. Lomuscio, Simon Parsons, Carles Sierra, and Michael J. Wooldridge, "Automated Negotiation: Prospects, Methods and Challenges," *International Journal of Group Decision and Negotiation* 10, no. 2 (2001): 199–215.

Katsh, M. Ethan, "Dispute Resolution in Cyberspace," *Connecticut Law Review* 28 (1995): 953.

Katsh, M. Ethan, and Orna Rabinovich-Einy, *Digital Justice: Technology and the Internet of Disputes* (Oxford: Oxford University Press, 2017).

Landsman, Stephan, "The Growing Challenge of Pro se Litigation," *Lewis & Clark Law Review* 13, no. 2 (2009): 439–460.

Lewis, Thomas Trent, "Helping Families by Maintaining a Strong Well-funded Family Court that Encourages Consensual Peacemaking: A Judicial Perspective," *Family Court Review* 53, no. 3 (2015): 371–377.

Lodder, Arno R., and John Zeleznikow, "Developing an Online Dispute Resolution Environment: Dialogue Tools and Negotiation Systems in a Three Step Model," *The Harvard Negotiation Law Review* 10 (2005): 287–338.

Lodder, Arno R., and John Zeleznikow, *Enhanced Dispute Resolution Through the Use of Information Technology* (Cambridge, New York: Cambridge University Press, 2010).

McCarty, L. Thorne, "Reflections on TAXMAN: An Experiment in Artificial Intelligence and Legal Reasoning," *Harvard Law Review* 90, no. 5 (1976): 837–893.

Nash, John, "Two-person Cooperative Games," *Econometrica: Journal of the Econometric Society* 21, no. 1 (1953): 128–140.

Raiffa, Howard, *The Art and Science of Negotiation* (Cambridge, MA: Harvard University Press, 1982).

Rule, Colin, *Online Dispute Resolution for Business: B2B, Ecommerce, Consumer, Employment, Insurance, and Other Commercial Conflicts* (San Francisco, CA: John Wiley & Sons, 2003).

Rule, Colin, and Larry Friedberg, "The Appropriate Role of Dispute Resolution in Building Trust Online," *Artificial Intelligence and Law* 13, no. 2 (2005): 193–205.

Salter, Shannon, and Darin Thompson, "Public-centered Civil Justice Redesign: A Case Study of the British Columbia Civil Resolution Tribunal," *McGill Journal of Dispute Resolution* 3 (2016): 113–136.

Sander, Frank E. A., "The Multi-door Courthouse: Settling Disputes in the Year 2000," *Barrister* 3 (1976): 18–42.

Schmitz, Amy J., and John Zeleznikow, "Intelligent Legal Tech to Empower Self-represented Litigants," *Columbia Science and Technology Law Review*, 23, no. 1 (2022): 142–190.

Smith, Roger, "Ministry of Justice for England and Wales Dives into the Deep Water on Online Dispute Resolution," *Dispute Resolution Magazine* 23 (2016): 28.

Sourdin, Tania, and John Zeleznikow, "Courts, Mediation and COVID-19," *Australian Business Law Review* 48, no. 2 (2020): 138–158.

Stranieri, Andrew, and John Zeleznikow, *Knowledge Discovery from Legal Databases*, Vol. 69 (Springer Science & Business Media, 2006).

Stranieri, Andrew, John Zeleznikow, Mark Gawler, and Bryn Lewis, "A Hybrid Rule–neural Approach for the Automation of Legal Reasoning in the Discretionary Domain of Family Law in Australia," *Artificial Intelligence and Law* 7, no. 2–3 (1999): 153–183.

Susskind, Richard E., *Expert Systems in Law* (Oxford: Clarendon Press, 1987).

Susskind, Richard E., *Online Courts and the Future of Justice* (Oxford: Oxford University Press, 2019).

Thiessen, Ernest, and John Zeleznikow, "Technical Aspects of Online Dispute Resolution Challenges and Opportunities," in *Proceedings of the Third Annual Forum on Online Dispute Resolution*, ed. Melissa Conley Tyler, Ethan Katsh, and Daewon Cho (Melbourne, 2004), 5–6.

Thomson, Mark, "Alternative Modes of Delivery for Family Dispute Resolution: The Telephone Dispute Resolution Service and the online FDR Project," *Journal of Family Studies* 17, no. 3 (2011): 253–257.

Toulmin, Stephen E., *The Uses of Argument* (Cambridge: Cambridge University Press, 1958).

Weizenbaum, Joseph, "ELIZA – A Computer Program for the Study of Natural Language Communication between Man and Machine," *Communications of the ACM* 9, no. 1 (1966): 36–45.

Wilson-Evered, Elisabeth, and John Zeleznikow, *Online Family Dispute Resolution: Evidence for Creating the Ideal People and Technology Interface* (Vol. 45) (New York: Springer Nature, 2021).

Wing, Leah, Janet Martinez, Ethan Katsh, Colin Rule, "Designing Ethical Online Dispute Resolution Systems: The Rise of the Fourth Party," *Negotiation Journal* 37, no. 1 (2021): 49–64.

Zeleznikow, John, "Using Web-based Legal Decision Support Systems to Improve Access to Justice," *Information and Communications Technology Law* 11, no. 1 (2002): 15–33.

Zeleznikow, John, "The Challenges of Using Online Dispute Resolution to Support Self Represented Litigants," *Journal of Internet Law* 23, no. 7 (2020): 3–14.

Zeleznikow, John, "Using Artificial Intelligence to provide Intelligent Dispute Resolution Support," *Group Decision and Negotiation* (2021): 1–24.

Zeleznikow, John, and Dan Hunter, *Building Intelligent Legal Information Systems: Representation and Reasoning in Law (No. 13)* (Deventer, Bosten, MA: Kluwer Law and Taxation Publishers, 1994).

Zeleznikow, John, and Andrew Stranieri, "The Split-up System: Integrating Neural Networks and Rule-based Reasoning in the Legal Domain," In *Proceedings of the 5th International Conference on Artificial Intelligence and Law* (New York: Association of Computing Machinery, 1995), 185–194.

CHAPTER 11

Total Surveillance – Everybody Watching Everybody Else

Vincent C. Müller

INTRODUCTION: PRIVACY, SURVEILLANCE, & MANIPULATION[1]

Privacy & Surveillance

There is a general discussion about privacy and surveillance in information technology, which mainly concerns access to private data and data that is personally identifiable.[2] Privacy has several well recognized aspects, e.g. "the right to be let alone," information privacy, privacy as an aspect of personhood, control over information about oneself, and the right to secrecy.[3] Privacy studies have historically focused on state surveillance by secret services but now include surveillance by other state agents, businesses, and even individuals. The technology has changed significantly in the last decades while regulation has been slow to respond. The result is a certain anarchy that is exploited by the most powerful players, sometimes in plain sight, sometimes in hiding.

The digital sphere has widened greatly: All data collection and storage is now digital, our lives are increasingly digital, most digital data is connected to a single Internet, and there is more and more sensor technology in use that generates data about non-digital aspects of our lives. AI increases both the possibilities of intelligent data collection and the possibilities for data analysis. This applies to blanket surveillance of whole populations as well as to classic targeted surveillance. In addition, much of the data is traded between agents, usually for a fee.

At the same time, controlling who collects which data, and who has access, is much harder in the digital world than it was in the analog world of paper and telephone calls. Many new AI technologies amplify the known issues. For example, face recognition in

DOI: 10.1201/9781003226406-12

photos and videos allows identification and thus profiling and searching for individuals.[4] This continues using other techniques for identification, e.g., "device fingerprinting," which are commonplace on the Internet (sometimes revealed in the "privacy policy"). The result is that "In this vast ocean of data, there is a frighteningly complete picture of us."[5] A scandal that still has not received due public attention.

The data trail we leave behind is how our "free" services are paid for – but we are not told about that data collection and the value of this new raw material, and we are manipulated into leaving ever more such data. For the "big 5" companies (Amazon, Google/Alphabet, Microsoft, Apple, Facebook), the main data collection part of their business appears to be based on deception, exploiting human weaknesses, furthering procrastination, generating addiction, and exercising manipulation.[6] The primary focus of social media, gaming, and most of the Internet in this "surveillance economy" is to gain, maintain and direct attention – and thus data supply. "Surveillance is the business model of the Internet."[7]

This surveillance and attention economy is sometimes called "surveillance capitalism."[8] It has caused many attempts to escape from the grasp of these corporations, e.g., in exercises of "minimalism,"[9] or through the open source movement, but it appears that present-day citizens have lost the degree of autonomy needed to escape while fully continuing with their life and work. We have lost ownership of our data, if "ownership" is the right relation here. It looks like we have lost control.

These systems will often reveal facts about us that we ourselves wish to suppress or are not aware of: they know more about us than we know ourselves. Even just observing online behavior allows insights into our mental states and manipulation (see below).[10] This has led to calls for the protection of "derived data."[11] With the last sentence of his bestselling book *Homo Deus*, Yuval Noah Harari asks about the long-term consequences of AI: "What will happen to society, politics and daily life when non-conscious but highly intelligent algorithms know us better than we know ourselves?"[12]

Robotic devices have not yet played a major role in this area, except for security patrolling, but this will change once they are more common outside of industry environments. Together with the "Internet of things," the so-called "smart" systems (phone, TV, oven, lamp, virtual assistant, home, …), the "smart city"[13] and "smart governance," they are set to become part of the data-gathering machinery that offers more detailed data, of different types, in real time, with ever more information.

Privacy-preserving techniques that can largely conceal the identity of persons or groups are now a standard staple in data science; they include (relative) anonymization, access control (plus encryption) and other models where computation is carried out with fully or partially encrypted input data[14]; in the case of "differential privacy" this is done by adding calibrated noise to encrypt the output of queries.[15] While requiring more effort and cost, such techniques can avoid many of the privacy issues. Some companies have also seen better privacy as a competitive advantage that can be leveraged and sold at a price.

One of the major practical difficulties is to actually enforce regulation, both on the level of the state and on the level of the individual who has a claim. They must identify the responsible legal entity, prove the action, perhaps prove intent, find a court that declares itself competent … and eventually get the court to actually enforce its decision. Well-established legal protections of consumer rights, product liability and other civil liability or protection of intellectual property rights are often missing in digital products or are hard to enforce. This means that companies with a "digital" background are used to testing their products on the consumers, without fear of liability, while heavily defending their intellectual property rights. This "Internet Libertarianism" is sometimes taken to assume that technical solutions will take care of societal problems by themselves.[16]

Manipulation of Behavior

The ethical issues of AI in surveillance go beyond the mere accumulation of data and direction of attention: They include the *use* of information to manipulate behavior, online and offline, in a way that undermines autonomous rational choice. Of course, efforts to manipulate behavior are ancient, but they may gain a new quality when they use AI systems. Given users' intense interaction with data systems, and the deep knowledge about individuals this provides, they are vulnerable to nudges, manipulation, and deception. With sufficient prior data, algorithms can be used to target individuals or small groups with just the kind of input that is likely to influence these particular individuals.

Many advertisers, marketers, and online sellers will use any legal means at their disposal, including exploitation of behavioral biases, deception, and the generation of addiction[17] – e.g., through "dark patterns" on web pages or in games.[18] Such manipulation is the business model in much of the gambling and gaming industries, but it is spreading, e.g., to low-cost airlines who make it difficult to book just a flight without adding travel insurance, a seat reservation, additional luggage, food, a rental car or a hotel room. Gambling and the sale of addictive substances are highly regulated, but online manipulation and addiction is not. Manipulation of online behavior is becoming a core business model of the Internet. Furthermore, social media are now the prime locations for political propaganda. This influence can be used to steer voting behavior, as in the Facebook-Cambridge Analytica scandal,[19] which may harm the autonomy of individuals if successful.[20]

Improved AI "faking" technologies make what once was reliable evidence into unreliable evidence. This has already happened to digital photos, sound recordings and video, and it will soon be quite easy to create (rather than alter) "deep fake" text, photos and video material with any content desired. Soon, sophisticated real-time interaction with persons over texting, phone or video will be faked, too. So, we cannot trust digital interaction, while we are at the same time increasingly dependent on it.

One more specific issue is that machine learning techniques in AI rely on training with vast amounts of data. This means there will often be a trade-off between privacy and rights to data versus technical quality of the product. This influences the consequentialist evaluation of privacy-violating practices.

The policy in this field has its ups and downs: Privacy protection has diminished massively as compared to the pre-digital age where communication was based on letters, analog telephone communications, and personal conversation – and surveillance operated under significant legal constraints. While the EU General Data Protection Regulation has strengthened privacy protection,[21] the United States and China prefer growth with less regulation,[22] likely in the hope that this provides a competitive advantage. Businesses, secret services, and other state agencies that depend on surveillance exert intense pressure to allow their operation and limit the protection of individual rights and civil liberties. It is clear that state and business actors have increased their ability to invade privacy and to manipulate people with the help of AI technology and will continue to do so to further their particular interests unless reined in by policy in the interest of general society.

PRIVACY FOR US, SURVEILLANCE FOR YOU

As we have seen, classic surveillance is top-down, from the more powerful to the less powerful. This top-down surveillance has increased in the past decades, largely due to technological developments in the increase in personal digital data generation (sensors + digital life), the growing ability to collect and store this data, and improved computer processing (hardware and algorithms) that can process this data and make it available in useful form for humans (e.g., in data visualization). Surveillance is widely accepted as the price of convenience – but also generates resistance and an emphasis on the ability or right to control information ourselves). However, both acceptance and resistance are highly culture-relative and in flux, and the results are typically paradoxical: We want surveillance for the others, but not for us. Researchers on a research project on the "smart home" will not have a smart home. Mark Zuckerberg covers up the camera on his laptop. And it is not just these usual suspects.

TOTAL TOP-DOWN SURVEILLANCE

Some people have used the term Data Surveillance, or Dataveillance in order to distinguish activities that are purposeful for surveillance (like following a person, or installing a "bug" in a room) from activities that are carried out for other purposes (like selling a mobile phone or controlling car tolls) but generate data that allows surveillance by machines.[23] The current culmination of dataveillance is an "Überveillance"[24] – we are now in a condition that used to be called "total information awareness" (see below) when it was still the state doing the surveillance[25] – in which effectively all data on everyone is under surveillance by someone.[26]

The detection and analysis of patterns in ever more data is now a business model that is used on a large scale in IT with services that are provided "free of charge." Why are they provided without an obvious payment? Because the payment is in the form of data. The data has value in several ways, but mainly because it allows to businesses to manipulate me into eventually parting with my money and buying things (from Amazon to "in-game purchases"). So, the surveillance is valuable because it enables the manipulation.

When companies like Google provide their services and control, connect their use to my ID (e-mail, phone number, name, address), they know my various devices, e-mail traffic, account use, logon patterns, search patterns, YouTube use, gaming use, cloud documents, friends and collaborators, photos and videos (geo-tagged), calendar, address book, location, etc. etc. … As a PBS documentary sums it up, "In this vast ocean of data, there is a frighteningly complete picture of us."[27]

In some ways Google knows more about me than any person does; it even knows many things about me that I don't know myself. Perhaps I am more politically right wing or more homosexual than I would like to admit? And, of course, I forget lots of details, but Google doesn't. And if I turn off my mobile phone when I leave the home, or I don't use a service … well, that's a pattern, too. And that is just one company: Add government sources that may be able to access more than just one company or add classical surveillance from sensing in homes and on the streets (microphones, cameras, motion-sensing, temperature, etc.), add private peer-to-peer surveillance and we are already pretty close to a total surveillance situation – without noticing much of it. My personal data have become a commodity that is accessed out of my control.

The ethical evaluation of this situation may just turn on the consequences and, in this situation, there is a (small) probability that the overall outcome may be more beneficial than a scenario in which it is transparent who has access to my personal data and where I can control that access. However, the current situation is problematic from a rights perspective since it involves deception and, crucially, it does not respect a right to privacy, in particular informational privacy, which is usually considered to be fundamental for a person's autonomy.[28]

TOTAL SURVEILLANCE

We have become used to the idea that all the traditional data, e.g., in banking and insurance or in state bureaucracy is now digital. But in addition to that b) many perhaps most of the activities in our lives are now digital, and thus produce digital data. (When did you last write correspondence on paper? If you are younger than 40 the answer may well be "never.") Last but not least, there is an increasing amount of sensors that turn analog life into digital data – e.g. your mobile phone is full of these (mine has at least: two cameras, two microphones, touch-sensitive screen, fingerprint sensor, GPS, compass, Wifi, bluetooth, 4G, cellular, near field communication, barometer, three-axis gyro, accelerometer, proximity sensor, ambient light sensor and "iBeacon microlocation"). More sensors are involved in all the 'smart' systems like "smart cities," "smart homes" (including "smart light bulbs," "smart floors," "smart doors"), "wearables," the "quantified self," electronic toys and the "Internet of Things."

Top-down surveillance in its various forms and aspects is the subject of classical surveillance and privacy studies. We want to take a different perspective. It is now sometimes suggested that the asymmetric information flow of classical surveillance can be countered by surveillance from the bottom up – sousveillance.[29] Sousveillance makes use of sensors on people, e.g., portable cameras and wearable computing (e.g. fitness trackers) to control state agents, to bring surveillance to our attention and assure freedom of information (in the same vein as "Wikileaks").[30] This is different from the recent trend to force state agents to wear recording devices themselves, e.g. police wearing body cameras. In sousveillance, e.g., the potential victims of police violence would wear the camera – and of course videos of police violence taken by victims of bystanders have caused much uproar, esp. in the United States.

Surveillance is becoming easier for everybody, for classical top-down surveillance (e.g., by state agents or businesses), for surveillance from the less powerful to the more powerful (sousveillance), and horizontally, between agents of similar power levels (institutional agents or people). Our working hypothesis is that surveillance and sousveillance will be joined by what we propose to call "horizontal surveillance," i.e., the surveillance from one agent to another on the same power level. This might be from one corporation to another, e.g., industrial espionage, or from one person to another. The three major factors which we mentioned above (data production, data access, data analysis) that have pushed surveillance and sousveillance will now also push horizontal surveillance. Just consider what you can find out about me, from just the Internet, what friends can see from the use of WhatsApp, Twitter, Facebook (are you "nearby"?), Google, location services, etc., etc.

Overall, we are heading toward a society of total surveillance of all by all, a panopticon in which not only the guards see the prisoners (surveillance), and the prisoners see the guard (sousveillance) but also the prisoners see the other prisoners (horizontal surveillance). The global village is becoming that village where pretty much everybody knows pretty much everything about everyone else except that, of course, it is not a village. We do not get the benefits of socially integrated village life; we only get the gossip, the control and the manipulation.

APPLICATION STUDY: SURVEILLANCE ON DATA WITHOUT INFORMATION?[31]

Background

In December 2005, US President George Bush admitted that he had ordered the National Security Agency (NSA), in the wake of the September 11th attacks, to "listen in" on communications between the United States and abroad, and also within the United States, without obtaining relevant court orders. The NSA had previously been responsible only for spying on foreign soil or on foreign spies within the United States.[32] This admission sparked a fairly intense debate about the justification of the move, and several prominent supporters of Bush's policy appeared in the media. In an appearance on *Hardball with Chris Matthews*, former Attorney General Edwin Meese said, among other things: "We are not talking about wiretapping. … This isn't eavesdropping, it is just surveillance." He further alluded that "certain technical procedures" were used, but that these are "classified."[33] Another former member of republican administrations, Charles Fried, asks what harm is done by such activity: "Is a person's privacy truly violated if his international communications are subject to this kind of impersonal, computerized screening?"[34] Judge Posner is explicit: "But machine collection and processing of data cannot, as such, invade privacy." What is argued here falls into a pattern that is increasingly popular recently: we just have machines, computers, analyzing data, nobody is listening in, nobody intrudes into your privacy, so don't worry, there is no problem. – Is this correct?

The Argument in Brief

I propose to investigate this question via an argument that runs as follows

1. The data of US citizens (and others) is systematically analyzed by data-mining techniques.

2. An analyzing machine either understands what it analyses or it does not.

3. If it does understand, privacy has been breached.

4. If it does not understand, the analysis must be completed by a human being, so privacy is breached.

5. Our privacy is breached by data-mining.

6. The breach of privacy in data-mining cannot be warranted by the results of data-mining themselves: If data-mining tags an individual as suspect, that tagging is not sufficient legal grounds for breaching the person's privacy.

Since this argument has a valid logical form (up to step (5) it is what is called a "constructive dilemma"), the only question is whether the three premises of this argument are true.

The argument above does not deal with the question which information should be private, to whom, and under what circumstances. It just clarifies whether the use of data-mining alters the privacy situation, in particular whether the use of machines could fail to violate privacy while conventional "reading" by a human would. It also argues (in premise 1) that this is not a theoretical problem, but actually occurs.

What is Known to be Done

Before we enter into some illustrative details, let it be clear that what is known to be known is very basic, and what is thought to be known is mostly speculation. The relevant activities are carried out by police, secret services and businesses. None of these want the public to know about their activities. Even much of the academic research in the area is classified. So, what we know are the very general official pronouncements and the occasional unintended news that surfaces.

It should be stressed, however, that in the United States there are many companies that deal with data that in the European Union would be considered private. This applies to all imaginable sorts of customer data, but also to state-collected records and the cross-referencing of these.[35] Specialized companies identify people's credit standing, name, address, phone numbers, height, weight and social security number.[36] Much of this potential has been used by police and secret services after the 9/11 attacks, esp. in the "Matrix" program.[37] All of it is used by private companies for all sorts of purposes.

Some of the major companies involved at present are Axicon, ChoicePoint and LexisNexis – the latter now owns Seisint and is part of UK-based Reed-Elsevier. ChoicePoint alone has about 17 billion public records, 250 TB of data. It acquires data worldwide. In 2001, ChoicePoint offered information to the US government about all South-American people, e.g. Mexican voters with name, address, ID no. and date of birth.[38] The offer was accepted. These companies work for all major corporations in the United States to support marketing, customer-service, product development, employee screening, risk assessment, access control, fraud-detection, etc. Some major software companies, such as Google, have their own data-mining techniques for Internet searchers, webmail users, etc. (some indications on http://www.google-watch.org/).

Privacy for Us, Surveillance for Them

Very little of this activity is the result of directed, legally controlled surveillance of particular people. Most of it is carried out by secret services on foreign nationals. If a state A spies on the nationals of state B, the laws of state A typically do not disallow this, and the danger of state B complaining is small, given that it has no evidence and is possibly doing the same to the nationals of state A. (In the case of Echelon, see below, it may well be that states A and B exchange information about the nationals of each, thus avoiding responsibility for spying on their own nationals.)[39]

In the following, I shall focus on the case of the United States. The official 9/11 commission acknowledges the difference between what is permitted domestically and what is permitted otherwise: "The FBI's job in the streets of the United States would thus be a domestic equivalent, operating under the U.S. Constitution and quite different laws and rules, to the jobs of the CIA's operative officers abroad." It continues, "The FBI is accustomed to carrying out sensitive intelligence collection operations in compliance with the law."[40] The Department of Defense TAPAC report proposes constraints based on privacy, but adds: "We recommend *excluding* from these requirements data mining that is limited to foreign intelligence that does not involve U.S. persons."[41]

According to official data, the United States operates some 15 different secret agencies.[42] Prominent for our purposes are the CIA, which "collects, analyzes and disseminates intelligence from all sources," the NSA, which "intercepts and analyzes foreign communications and breaks code,"[43] the Geospatial-Intelligence Agency (GSA), which "provides and analyzes imagery and produces a wide array of products, including maps, navigation tools, and surveillance intelligence," and finally the National Reconnaissance Office (NRO), which "procures, launches and maintains in orbit information gathering satellites that serve other government agencies."[44]

Given the secrecy and power of such agencies, one might get the impression that they know everything and can do anything – but we should be careful not to give in to this temptation. There are many well documented failures where US agencies did now know about important events before they occurred: e.g., the fall of the Berlin Wall 1989, nuclear weapon tests by India and Pakistan 1998, attacks on New York and Langley 2001;[45] they also failed to identify the whereabouts of Saddam Hussein in 2002, or of Osama Bin Laden and other Al Qaeda operatives in recent years.

Surveillance and Terrorism

With the waning of the cold war and particularly after the attacks on September 11, 2001, the perception in the United States as to who are the potential enemies that should be observed has changed significantly. The target is now "terrorism," which is far less specific than particular foreign states or foreign military – but still couched in the same terms as a "war." It involves groups or even individuals who may well not have identified themselves as enemies before they attack. Accordingly, the techniques for identifying such enemies, preventing their attacks and persecuting perpetrators would have to adjust.[46] One cannot expect few, rich sources of information, as may be the case in traditional espionage, but has to make sense of many small pieces of information. Having said that, the NSA publicly admitted that before 9/11 "FBI headquarters routinely receive about 200 reports daily from us." But adds "... we have been more agile in sharing information with some customers (like the Department of Defense) than we have with others (like the Department of Justice)."[47] There was supposed to be information sharing from the FBI to the intelligence services, but not in the opposite direction. This obstacle was frequently blamed for the failure to identify 9/11 (sometimes called "the wall") and US Congress suggested "... develop[ing] capability to facilitate the timely and complete sharing of relevant intelligence information both within the Intelligence Community and with other appropriate federal, state, and local authorities."[48]

Allow me to mention some known programs:

In 2002, the US government proposed TIA, the "Total Information Awareness" system, in which all US bank records, tax filings, driver license information, credit card information, financial records, medical data, telephone and e-mail records, travel information and other data would be combined under the auspice of DARPA – who had been working on the concept since the late 1990s.[49] These databases were to be augmented with conventional intelligence, improved human identification, speech-to-text (EARS: effective affordable reusable speech to text), automatic translation technologies (TIDES) and analyzed with data-mining technologies (in GENISYS). In one 2002 presentation, the "automation goal" was described as "read everything without reading everything."[50] TIA is deeply embedded into various DARPA computing technology projects.[51] Following widespread criticism, the program was renamed "Terrorism Information Awareness" in March 2003 but had to be dropped due to the continuing criticism and a forthcoming damning report of the Department of Defense.[52] The order to cancel such activities was signed by President Bush in October 2003, but contained a "classified annex," specifying for which purposes the continuation of the program was permitted.[53] What was shocking to the US public was the use of private US data. However, what has clearly continued is the use of private data from abroad and from non-US citizens. This includes data gathered by security services in violation of local law.

A similar program was *"Matrix"* (Multi-state Anti-Terrorism Information Exchange), which aimed to combine these publicly available identifiable records and, when joined by the FBI, added criminal records, driver license photos – and all that through a private company with experience in commercial data-mining: Hank Asher's *Seisint*.[54] Though police were impressed with the new abilities of this "one-stop shop," public concern about this combination of criminal and commercial records forced the cancellation of US-wide Matrix late in 2003 when more and more US states refused to cooperate. However, some US states still appear to work with the system.[55] O'Harrow says, "Make no mistake, though, Asher's technology will be there, working behind the scenes, no matter what it is called, and now quite possibly on a global scale."[56] Bailey deplores the cancellation of TIA, but also offers this consolation, "Going forward, it's likely that many of the ideas and much of the technology behind TIA will live on and be utilized by different agencies, [...]."[57] Given the

"Classified Annex" mentioned above, this is not just speculation, but certainty. Also, the founding of the National Counterterrorism Center may point in this direction.

The 2001 Patriot Act required many companies and organizations in the United States to report "suspicious activity." The *Financial Crimes Enforcement Network*, FinCEN, is now used by banks to report "suspicious" customer behavior to the FBI – about 300.000 reports were filed in 2003 alone. FinCEN also shares its entire database with the FBI each month. Customers are not informed.[58]

Echelon is a network of secret services operated by Australia, Canada, New Zealand, the United Kingdom and the United States, which is thought allow near-total surveillance of communication by potentially intercepting all wireless communications worldwide. The Soviet Union had a similar network, called Dozor, which is currently being updated.[59]

Everyone who boards a flight to or inside the United States has been checked by "Computer Assisted Passenger Pre-Screening" (CAPPS), installed in the late 1990s and updated in 2003 to CAPPS II, run by the Transportation Safety Agency (TSA), a division of the Department of Homeland Security. Any passenger's name, address, US address, date of birth and US phone number is required. TSA will query various databases and classify passengers as "green," "yellow" (special scrutiny) or "red" (inform police). CAPPS does not mine data, it matches entries from many databases it queries.[60]

The FBI has a system called *"Carnivore"* which it installs in the computers of Internet service providers. The system reads the headers of all e-mails sent and received, as well as the IP addresses of all communications, such as web sites, ftp sites, etc. If a communication matches the search criteria, its content is stored and sent to the FBI. It is supposed to catch only information relevant to a particular suspect, and to be subject to the authorization by a court order, including the secret FISA courts.[61] Similar systems are in use in other countries as well.

Given this quick survey of known surveillance activities by US services and businesses, it seems that we have established our first premise, that data-mining on our data is actually taking place – on communications, databases and surveillance data. I presume that it is granted that the machines either understand or do not understand what they analyze, even if there may be a gray zone here, so what we need to see now is whether this constitutes a violation of privacy.

Machine Understanding

To be brief, it is almost universally agreed that current computer systems do not understand, states that are *about* something in the world – they lack *intentional states*.[62] So, when they find a particular pattern, they cannot know what that pattern means. John Searle has presented an influential argument that *no* computer will be able to understand, at least not by virtue of being a computer.[63] This raises the question whether our fourth premise is correct

> (4) If it does not understand, the analysis must be completed by a human being, so privacy is breached.

How far will an automated analysis without understanding go? Will statistics distinguish the guilty from the suspicious (the terrorist from the philosopher who analyzes problems of privacy)? The experts' verdict is quite clear. DeRosa summarizes her view: "These techniques ... are not likely to be useful as the only source for a conclusion or decision."[64] Taipale, generally a supporter of data-mining, says that "a guiding principle ... should be that data mining not be used to automatically trigger law enforcement consequences."[65] It lies in the nature of data-mining for security, that it will only produce a certain probability for a match, but also many false positives. Unless a human investigator is involved at this stage, who views the data, the analysis remains useless. But at this stage the fact that someone is considered "suspect" by the system does not mean he/she is sufficiently suspect to warrant a breach of privacy, e.g., a court order.

Reports in the press indicate that problem of false positives is currently acute: *The Washington Post* headlined "325,000 Names on Terrorism List: Rights Groups Say Database May Include Innocent People."[66] Officials from the National Counterterrorism Center (NCTC) responded by pointing out that a number of people may be entered with several names and spellings, thus reducing the number to more than 200,000 actual people. Some 32,000 people on this list have been coded as "armed and dangerous" but have been given "the lowest handling code," meaning FBI is not informed if they are encountered. In other words, suspicions are not substantial enough. The Washington Post picked up neither on this self-confessed indication of incompetence (1/3 of names are misleading) nor on the obvious irony of the phrase "may include innocent people" in the title. Ignoring that most almost all suspects are innocent is a large part of the political problem.

A Small Price to Pay?

Now, even if we accept premise (4), this does not imply that anything untoward is happening. As we said above, it is clear that there are circumstances under which the right to privacy should be breached for a particular person. So, if the machine pattern is enough for a serious suspicion as defined in the relevant law, then a court order should be obtainable as implied by premise (6) above. Clearly, one must be aware that total privacy never existed and that breach of privacy is desirable, even for the totally innocent. Good cases for this are made, in some detail, in the work of Bailay and Etzioni.[67]

One other important aspect is how much we trust the agency that is breaching privacy. If it is run by a totalitarian state or a profit-oriented business, we will assume that its purposes are at least partially problematic – but we cannot assume the same for a democratically controlled state structure. This is one of the issues that make the discussion so difficult: in order to evaluate privacy breaches, one may think one has to evaluate the motives of the agencies concerned; and little agreement will be reached on this point. It is clear that malicious agencies can do nasty things, and it is also clear that agencies will not admit to any malicious intentions, so I suggest that we try to prove our point even under the assumption of a totally benevolent agency. Note that doing this will imply the demand for significantly reduced rights to profit-oriented businesses who carry out data-mining.

CONCLUSION: WHAT SHOULD NOT BE DONE INNOCENTLY

What we have established here is that the use of data-mining techniques constitutes a breach of privacy as soon as a person is involved. That breach is justified, if and only if that person is legally authorized to breach this privacy in this case. Given that data-mining for security purposes without the involvement of persons is currently impossible, its use always constitutes a breach of privacy. Given that current data-mining is too weak to establish legally sufficient grounds for a violation of privacy, it is ethically wrong.

The more sophisticated defenders of data mining suggest that its problem lies in *how* it is used: "One of the principal reasons for public concern about these tools is that there appears to be no consistent policy guiding decisions when and how to use them."[68] I have tried to show that *any* use violates a right to privacy unless it is authorized by the person concerned. Subject-based analysis and matching already presuppose personally identified data, while pattern analysis requires human intervention. All use of blanket data analysis should therefore be subject to the conventional legal controls, which excludes using them on whole populations. We should try to make sure that at least our own state and private agencies do not cross these lines.

ACKNOWLEDGMENTS

Some of the ideas around total surveillance were developed in conversations with Stuart Armstrong when we shared an office at FHI in Littlegate House, Oxford. I am grateful for the reviewer for CPG Publications for pointing out several little issues that needed improvement.

NOTES

1 This section is based on my article "Ethics of Artificial Intelligence and Robotics," in *Stanford Encyclopedia of Philosophy*, ed. Edward N. Zalta (Palo Alto: CSLI, Stanford University, 2020), 1–70, https://plato.stanford.edu/entries/ethics-ai/.

2 See e.g., Kevin Macnish, *The Ethics of Surveillance: An Introduction* (London: Routledge, 2017); Beate Roessler, "Privacy as a Human Right," *Proceedings of the Aristotelian Society* 2 (CXVII) (2017).

3 Coilin J. Bennett, and Charles Raab, *The Governance of Privacy: Policy Instruments in Global Perspective* (3rd ed., Cambridge, MA: MIT Press, 2017).

4 Alex Campolo, Madelyn Sanfilippo, Meredith Whittaker, and Kate Crawford, "AI Now Report 2017," 2017, 15 ff., https://ainowinstitute.org/AI_Now_2017_Report.html.

5 Sandy Smolan, "The Human Face of Big Data," PBS Documentary, February 24, 2016, 1:01, https://www.youtube.com/watch?v=kAZ8lK224Kw.

6 Tristan Harris, "How Technology is Hijacking Your Mind – From a Magician and Google Design Ethicist," *Thrive Global*, May 19, 2016, https://medium.com/thrive-global/how-technology-hijacks-peoples-minds-from-a-magician-and-google-s-design-ethicist-56d62ef5edf3.

7 Bruce Schneier, *Data and Goliath: The hidden battles to collect your data and control your world* (New York: W. W. Norton, 2015).

8 Shoshana Zuboff, *The Age of Surveillance Capitalism: The Fight for a Human Future at the New Frontier of Power* (New York: Public Affairs, 2019).

9 Cal Newport, *Digital Minimalism: On Living Better with Less Technology* (London: Penguin, 2019).

10 Christopher Burr, and Nello Christianini, "Can Machines Read Our Minds?" *Minds and Machines* 29, no. 3 (2019): 461–494.

11 Sandra Wachter and Brent Daniel Mittelstadt, "A Right to Reasonable Inferences: Re-thinking Data Protection Law in the Age of Big Data and AI," *Columbia Business Law Review*, no. 2 (2019): 494–620.

12 Yuval Noah Harari, *Homo Deus: A Brief History of Tomorrow* (New York: Harper, 2016).

13 Richard Sennett, *Building and Dwelling: Ethics for the City* (London: Allen Lane, 2018).

14 Bernd Carsten Stahl, and David Wright, "Ethics and Privacy in AI and Big Data: Implementing Responsible Research and Innovation," *IEEE Security & Privacy* 16, no. 3 (2018): 26–33.

15 John M. Abowd, "How will Statistical Agencies Operate When All Data are Private?" *Journal of Privacy and Confidentiality* 7, no. 3 (2017): 1–15; Cynthia Dwork et al., "Calibrating Noise to Sensitivity in Private Data Analysis," in *Theory of Cryptography. TCC 2006. Lecture Notes in Computer Science*, Vol. 3876, ed. Shai Halevi and Tal Rabin (Berlin, Heidelberg: Springer, 2006), 265–284.

16 Eygeny Mozorov, *To Save Everything, Click Here: The Folly of Technological Solutionism* (New York: Public Affairs, 2013).

17 Elisabeth Costa and David Halpern, "The Behavioural Science of Online Harm and Manipulation, and What to do About It: An Exploratory Paper to Spark Ideas and Debate," *The Behavioural Insights Team Report*, 2019, 1–82, https://www.bi.team/publications/the-behavioural-science-of-online-harm-and-manipulation-and-what-to-do-about-it/.

18 Arunesh Mathur et al., "Dark Patterns at Scale: Findings from a Crawl of 11K Shopping Websites," *Proceedings of the ACM Human-Computer Interaction* 3, no. 81 (2109): 1–32.

19 Samantha Bradshaw, Lisa-Maria Neudert, and Phil Howard, "Government Responses to Malicious Use of Social Media," Oxford Project on Computational Propaganda, Working Paper 2019.2., 2019, https://comprop.oii.ox.ac.uk/research/government-responses/; Sam Woolley and Phil Howard (eds.), *Computational Propaganda: Political Parties, Politicians, and Political Manipulation on Social Media* (Oxford: Oxford University Press, 2017).

20 Daniel Susser, Beate Roessler, and Helen Nissenbaum, "Technology, Autonomy, and Manipulation," *Internet Policy Review* 8, no. 2 (2019): 1–22.

21 "General Data Protection Regulation: Regulation (EU) 2016/679 of the European Parliament and of the Council of 27 April 2016 on the protection of natural persons with regard to the processing of personal data and on the free movement of such data, and repealing Directive 95/46/EC," *Official Journal of the European Union* 119 (4 May 2016), 1–88, http://data.europa.eu/eli/reg/2016/679/oj.

22 Nicholas Thompson and Ian Bremmer, "The AI Cold War that Threatens Us All," *Wired*, October 23, 2018, https://www.wired.com/story/ai-cold-war-china-could-doom-us-all/.

23 Vincent C. Müller, "Would You Mind Being Watched by Machines? Privacy Concerns in Data Mining," *AI & Society* 23, no. 4 (2009): 529–544.

24 Richard Clarke, "What is Überveillance? (And What Should be Done About it?)," *IEEE Technology and Society Magazine* 29, no. 2 (2010): 17–25.

25 Jack M. Balkin, "The Constitution in the National Surveillance State," *Minnesota Law Review* 93, no. 1 (2008): 1–25.

26 Bruce Schneier, *Data and Goliath: The hidden battles to collect your data and control your world* (New York: W. W. Norton, 2015).

27 Sandy Smolan, "The Human Face of Big Data," PBS Documentary, February 24, 2016, 1:01, https://www.youtube.com/watch?v=kAZ8lK224Kw.

28 Beate Roessler, "Privacy as a Human Right," *Proceedings of the Aristotelian Society* 2 (CXVII) (2017).

29 Stephen Mann, "Veilance and Reciprocal Transparency: Surveillance versus Sousveillance, AR Glass, Lifeglogging, and Wearable Computing," in *2013 IEEE International Symposium on Technology and Society (ISTAS): Social Implications of Wearable Computing and Augmediated Reality in Everyday Life*, 2013, 1–12.

30 Ethan Zuckermann, "Die Antwort auf Überwachung heisst Unterwachung," *Die Zeit*, October 7, 2013.

31 This section is based on my earlier paper "From Public Data to Private Information: The Case of the Supermarket," in *Proceedings of the 8th International Conference Computer Ethics: Philosophical Enquiry*, ed. Maria Bottis (Corfu: Nomiki Bibliothiki, 2009), 500–507, https://philpapers.org/rec/MLLFPD-2.

32 "The Law Requires the NSA to not Deliberately Collect Data on US Citizens or on Persons in the United States without a Warrant Based on Foreign Intelligence Requirements." (9/11 Commission 2004, 87) This avoidance of domestic data was considered a significant factor for the failure to prevent the September 11th 2001 attacks ("From Public Data to Private Information: The Case of the Supermarket," in *Proceedings of the 8th International Conference Computer Ethics: Philosophical Enquiry*, ed. Maria Bottis (Corfu: Nomiki Bibliothiki, 2009), 500–507).

33 Edwin Meese on MSNBC, in Chris Matthews' program "Hardball", Jan 12th, 2006, 19:45.

34 Fried in his article "The Case for Surveillance," *The Boston Globe*, December 30, 2005 (updated). Fried was solicitor general in the second Reagan administration. Taipale argues data-mining is "…different than claiming that 'everybody is being investigated' through pattern-matching. In reality only the electronic footprints of transactions and activities are being scrutinized." Kim A. Taipale, "Data Mining and Domestic Security: Connecting the Dots to Make Sense of Data," *Columbia Science and Technology Law Review* 5, no. 2 (2003): 66.

35 See Markle Foundation, "Creating a Trusted Network for Homeland Security: A Report of the Markle Foundation Task Force," 2003, http://www.markletaskforce.org.

36 Mary DeRosa, *Data Mining and Data Analysis for Counterterrorism* (Center for Strategic and International Studies report; Washington: CSIS Press, 2004), 10.

37 Robert O'Harrow, *No Place to Hide: Behind the Scenes of our Emerging Surveillance Society* (New York: Simon & Schuster, 2005), Chapters 2 and 4.

38 Robert O'Harrow, *No Place to Hide: Behind the Scenes of our Emerging Surveillance Society* (New York: Simon & Schuster, 2005), 145, 152.

39 Kevin Keenan, *Invasion of Privacy: A Reference Handbook* (Santa Barbara: ABC-CLIO, 2005), 43.

40 "The Law Requires the NSA to not Deliberately Collect Data on US Citizens or on Persons in the United States Without a Warrant Based on Foreign Intelligence Requirements." (9/11 Commission 2004, 423).

41 Their italics, TAPAC, "Safeguarding Privacy in the Fight against Terrorism," Report of the Technology and Privacy Advisory Committee to the Department of Defense, March 1, 2004, x, http://purl.access.gpo.gov/GPO/LPS52114. The same provisions, excluding military operations and intelligence activities overseas or against non-US citizens are in the "Department of Homeland Securities Appropriations Act," sect. 8131 (b), as quoted in Kim A. Taipale, "Data Mining and Domestic Security: Connecting the Dots to Make Sense of Data," *Columbia Science and Technology Law Review* 5, no. 2 (2003), 10, Beate Roessler, "Privacy as a Human Right," *Proceedings of the Aristotelian Society* 2 (CXVII) (2017)).

42 The Law Requires the NSA to not Deliberately Collect Data on US Citizens or on Persons in the United States Without a Warrant Based on Foreign Intelligence Requirements." (9/11 Commission 2004, 410).

43 For more information about the NSA, consider (Anonymous 2005).

44 NSA, consider (Anonymous 2005), 86 f.

45 About the failure to identify September 11th: "To put this into perspective, throughout the summer of 2001 we had more than 30 warnings that *something* was imminent. We dutifully reported these, yet none of these subsequently correlated with terrorist attacks. The concept of 'imminent' to our adversaries is relative; it can mean soon or simply sometime in the future." (Michael V. Hayden, "Statement for the Record by Lieutenant General Michael V. Hayden, USAF, Director, National Security Agency/Chief, Central Security Service, before the Joint Inquiry of the Senate Select Committee on Intelligence and the House Permanent Select Committee on Intelligence, 17 October 2002," 2002, 4, http://www.gwu.edu/~nsarchiv/NSAEBB/NSAEBB24/nsa27.pdf). Hayden also stresses the difficulty of identifying and processing several languages and the crucial factor of processing on time.

46 Markle Foundation, "Protecting America's Freedom in the Information Age: A Report of the Markle Foundation Task Force," 2002, http://www.markletaskforce.org.

47 Michael V. Hayden, "Statement for the Record by Lieutenant General Michael V. Hayden, USAF, Director, National Security Agency/Chief, Central Security Service, before the Joint Inquiry of the Senate Select Committee on Intelligence and the House Permanent Select Committee on Intelligence, 17 October 2002," 2002, 9.

48 Quoted in Kim A. Taipale, "Data Mining and Domestic Security: Connecting the Dots to Make Sense of Data," *Columbia Science and Technology Law Review* 5, no. 2 (2003): 5.

49 Robert O'Harrow, *No Place to Hide: Behind the Scenes of our Emerging Surveillance Society* (New York: Simon & Schuster, 2005), Chapter 7; Richard A. Rosenberg (ed.), *The Social Impact of Computers* (3rd ed., San Diego: Elsevier, 2004), 390 f.

50 Tom Armour, "Genoa II and Total Information Awareness," 2002 DARPATECH Symposium "Transforming Fantasy," 2002, http://www.darpa.mil/darpatech2002/presentations/iao_pdf/slides /armouriao.pdf.

51 Defense Advanced Research Projects Agency (DARPA), "Fact File: A Compendium of DARPA Programs," August 2003 (updated), 5–11, http://www.darpa.mil/body/news/2003/final2003factfilerev1.pdf; Information Awareness Office (Department of Defense), "Report to Congress Regarding the Terrorism Information Awareness Program," May 20, 2003 (updated), http://wyden.senate.gov/leg_issues/reports/darpa_tia_summary.pdf.

52 TAPAC, "Safeguarding Privacy in the Fight against Terrorism," Report of the Technology and Privacy Advisory Committee to the Department of Defense, March 1, 2004, x, http://purl.access.gpo.gov/GPO/LPS52114. Most references to TIA have been removed from the DARPA sites, but Director Pointexter's outlook can be gathered from his slides. See Poindexter, John, "Information Awareness Office Overview," Introductory Statement to the 2002 DARPATECH symposium, 2002, http://www.darpa.mil/darpatech2002/presentations/iao_pdf/ speeches/poindext.pdf.

53 Kim A. Taipale, "Data Mining and Domestic Security: Connecting the Dots to Make Sense of Data," *Columbia Science and Technology Law Review* 5, no. 2 (2003): 10; cf. 39 ff.

54 Robert O'Harrow, *No Place to Hide: Behind the Scenes of our Emerging Surveillance Society* (New York: Simon & Schuster, 2005), 107 f., 121 ff.

55 Kim A. Taipale, "Data Mining and Domestic Security: Connecting the Dots to Make Sense of Data," *Columbia Science and Technology Law Review* 5, no. 2 (2003): 16, still hopes for its nationwide extension.

56 Robert O'Harrow, *No Place to Hide: Behind the Scenes of our Emerging Surveillance Society* (New York: Simon & Schuster, 2005), 124.

57 Dennis Bailey, *The Open Society Paradox: Why the Twenty-first Century Calls for More Openness – not Less* (Washington: Brasseys, 2004), 106.

58 Robert O'Harrow, *No Place to Hide: Behind the Scenes of our Emerging Surveillance Society* (New York: Simon & Schuster, 2005), 266.

59 Kevin Keenan, *Invasion of Privacy: A Reference Handbook* (Santa Barbara: ABC-CLIO, 2005), 42 f.

60 Kim A. Taipale, "Data Mining and Domestic Security: Connecting the Dots to Make Sense of Data," *Columbia Science and Technology Law Review* 5, no. 2 (2003), 37 ff.

61 Kevin Keenan, *Invasion of Privacy: A Reference Handbook* (Santa Barbara: ABC-CLIO, 2005), 71; Robert O'Harrow, *No Place to Hide: Behind the Scenes of our Emerging Surveillance Society* (New York: Simon & Schuster, 2005), 257.

62 For an introduction, see Tim Crane, *The Mechanical Mind: A Philosophical Introduction to Minds, Machines and Mental Representation* (2nd ed., London: Routledge 2003), esp. Chapter 3.

63 Cf. John Preston, and Mark Bishop (eds.), *Views into the Chinese Room: New Essays on Searle and Artificial Intelligence* (Oxford: Oxford University Press, 2002); John R. Searle, "Minds, Brains and Programs," *Behavioral and Brain Sciences* 3, no. 3 (1980): 417–457.

64 Mary DeRosa, *Data Mining and Data Analysis for Counterterrorism* (Center for Strategic and International Studies report; Washington: CSIS Press, 2004), v.

65 Kim A. Taipale, "Data Mining and Domestic Security: Connecting the Dots to Make Sense of Data," *Columbia Science and Technology Law Review* 5, no. 2 (2003), 32.

66 Walter Pincus Dan Eggen, "325,000 Names on Terrorism List: Rights Groups Say Database May Include Innocent People," *The Washington Post*, February 15, 2006 (updated), http://www.washingtonpost.com/wp-dyn/content/article/2006/02/14/AR2006021402125.html.

67 Dennis Bailey, *The Open Society Paradox: Why the Twenty-first Century Calls for More Openness – not Less* (Washington: Brasseys, 2004), Part III; Amitai Etzioni, *The Limits of Privacy* (New York: Basic Books, 1999).

68 Mary DeRosa, *Data Mining and Data Analysis for Counterterrorism* (Center for Strategic and International Studies report; Washington: CSIS Press, 2004), vii.

REFERENCES

9/11 Commission, *The Final Report of the National Commission on Terrorist Attacks Upon the United States* (Commission instituted November 27th, 2002 by US President and Congress, public hearings March 2003–June 2004; New York: Norton, 2004).

Abowd, John M., "How will Statistical Agencies Operate When All Data are Private?" *Journal of Privacy and Confidentiality* 7, no. 3 (2017): 1–15.

Anonymous, "National Security Archive," 2005, https://www.gwu.edu/~nsarchiv/NSAEBB/NSAEBB24/index.htm.

Armour, Tom, "Genoa II and Total Information Awareness,"2002 DARPATECH Symposium "Transforming Fantasy," 2002, https://www.darpa.mil/darpatech2002/presentations/iao_pdf/slides /armouriao.pdf.

Bailey, Dennis, *The Open Society Paradox: Why the Twenty-first Century Calls for More Openness – not Less* (Washington: Brasseys, 2004).

Balkin, Jack M., "The Constitution in the National Surveillance State," *Minnesota Law Review* 93, no. 1 (2008): 1–25.

Bennett, Colin J., and Charles Raab, *The Governance of Privacy: Policy Instruments in Global Perspective* (3rd ed., Cambridge, MA: MIT Press, 2017).

Bradshaw, Samantha, Lisa-Maria Neudert, and Phil Howard, "Government Responses to Malicious Use of Social Media," Oxford Project on Computational Propaganda, Working Paper 2019.2., 2019, https://comprop.oii.ox.ac.uk/research/government-responses/.

Burr, Christopher, and Nello Christianini, "Can Machines Read Our Minds?" *Minds and Machines* 29, no. 3 (2019): 461–494.

Campolo, Alex, Madelyn Sanfilippo, Meredith Whittaker, and Kate Crawford, "AI Now Report 2017," 2017, https://ainowinstitute.org/AI_Now_2017_Report.html.

Clarke, Richard, "What is Überveillance? (And What Should be Done About it?)," *IEEE Technology and Society Magazine* 29, no. 2 (2010): 17–25.

Costa, Elisabeth, and David Halpern, "The Behavioural Science of Online Harm and Manipulation, and What to do About It: An Exploratory Paper to Spark Ideas and Debate," The Behavioural Insights Team Report, 2019, 1–82, https://www.bi.team/publications/the-behavioural-science-of-online-harm-and-manipulation-and-what-to-do-about-it/.

Crane, Tim, *The Mechanical Mind: A Philosophical Introduction to Minds, Machines and Mental Representation* (2nd ed., London: Routledge 2003).

Defense Advanced Research Projects Agency (DARPA), "Fact file: A Compendium of DARPA Programs," August 2003 (updated), https://www.darpa.mil/body/news/2003/final2003 factfilerev1.pdf.

DeRosa, Mary, *Data Mining and Data Analysis for Counterterrorism* (Center for Strategic and International Studies report; Washington: CSIS Press, 2004).

Dwork, Cynthia, Frank McSherry, Kobbi Nissim, and Adam Smith, Adam, "Calibrating Noise to Sensitivity in Private Data Analysis," in *Theory of Cryptography. TCC 2006. Lecture Notes in Computer Science, Vol. 3876*, ed. Shai Halevi and Tal Rabin (Berlin, Heidelberg: Springer, 2006), 265–284.

Etzioni, Amitai, *The Limits of Privacy* (New York: Basic Books, 1999).

Fried, Charles, "The Case for Surveillance," *The Boston Globe*, December 30, 2005 (updated), https://www.boston.com/news/globe/editorial_opinion/oped/articles/2005/12/30/the_case_for_surveillance?mode=PF.

GDPR, "General Data Protection Regulation: Regulation (EU) 2016/679 of the European Parliament and of the Council of 27 April 2016 on the Protection of Natural Persons with Regard to the Processing of Personal Data and on the Free Movement of Such Data, and Repealing Directive 95/46/EC," *Official Journal of the European Union* 119 (4 May 2016), 1–88, https://data.europa.eu/eli/reg/2016/679/oj.

Harari, Yuval Noah, *Homo Deus: A Brief History of Tomorrow* (New York: Harper, 2016).

Harris, Tristan, "How Technology is Hijacking Your Mind – From a Magician and Google Design Ethicist," *Thrive Global*, May 19, 2016, https://medium.com/thrive-global/how-technology-hijacks-peoples-minds-from-a-magician-and-google-s-design-ethicist-56d62ef5edf3.

Hayden, Michael V., "Statement for the Record by Lieutenant General Michael V. Hayden, USAF, Director, National Security Agency/Chief, Central Security Service, before the Joint Inquiry of the Senate Select Committee on Intelligence and the House Permanent Select Committee on Intelligence, 17 October 2002," https://www.gwu.edu/~nsarchiv/NSAEBB/NSAEBB24/nsa27.pdf.

Information Awareness Office (Department of Defense), "Report to Congress Regarding the Terrorism Information Awareness Program," May 20, 2003 (updated), https://wyden.senate.gov/leg_issues/reports/darpa_tia_summary.pdf.

Keenan, Kevin, *Invasion of Privacy: A Reference Handbook* (Santa Barbara: ABC-CLIO, 2005).

Macnish, Kevin, *The Ethics of Surveillance: An Introduction* (London: Routledge, 2017).

Mann, Stephen, "Veilance and Reciprocal Transparency: Surveillance versus Sousveillance, AR Glass, Lifeglogging, and Wearable Computing," 2013 IEEE International Symposium on Technology and Society (ISTAS): Social Implications of Wearable Computing and Augmediated Reality in Everyday Life, 2013, 1–12.

Markle Foundation, "Protecting America's Freedom in the Information Age: A Report of the Markle Foundation Task Force," 2002, https://www.markletaskforce.org.

Markle Foundation, "Creating a Trusted Network for Homeland Security: A Report of the Markle Foundation Task Force," 2003, https://www.markletaskforce.org.

Mathur, Arunesh, Gunes Acar, Michael Friedman, Elena Lucherini, Jonathan Mayer, Marshini Chetty, and Arvind Narayanan, "Dark Patterns at Scale: Findings from a Crawl of 11K Shopping Websites," *Proceedings of the ACM Human-Computer Interaction* 3, no. 81 (2109): 1–32.

Mozorov, Eygeny, *To Save Everything, Click Here: The Folly of Technological Solutionism* (New York: Public Affairs, 2013).

Müller, Vincent C., "Would You Mind being Watched by Machines? Privacy Concerns in Data Mining," *AI & Society* 23, no. 4 (2009): 529–544.

Müller, Vincent C., "From Public Data to Private Information: The Case of the Supermarket," in *Proceedings of the 8th International Conference Computer Ethics: Philosophical Enquiry*, ed. Maria Bottis (Corfu: Nomiki Bibliothiki, 2009), 500–507, https://philpapers.org/rec/MLLFPD-2.

Müller, Vincent C., "Ethics of Artificial Intelligence and Robotics," In *Stanford Encyclopedia of Philosophy*, ed. Edward N. Zalta (Palo Alto: CSLI, Stanford University, 2020), 1–70, https://plato.stanford.edu/entries/ethics-ai/.

Newport, Cal, *Digital Minimalism: On Living Better with Less Technology* (London: Penguin, 2019).

O'Harrow, Robert, *No Place to Hide: Behind the Scenes of Our Emerging Surveillance Society* (New York: Simon & Schuster, 2005).

Pincus, Walter, and Dan Eggen, "325,000 Names on Terrorism List: Rights Groups Say Database May Include Innocent People," *The Washington Post*, February 15, 2006 (updated), https://www.washingtonpost.com/wp-dyn/content/article/2006/02/14/AR2006021402125.html.

Poindexter, John, "Information Awareness Office Overview," Introductory Statement to the 2002 DARPATECH Symposium, 2002, https://www.darpa.mil/darpatech2002/presentations/iao_pdf/ speeches/poindext.pdf.

Preston, John, and Mark Bishop (eds.), *Views into the Chinese Room: New Essays on Searle and Artificial Intelligence* (Oxford: Oxford University Press, 2002).

Roessler, Beate, "Privacy as a Human Right," *Proceedings of the Aristotelian Society* 2 (CXVII) (2017).

Rosenberg, Richard A. (ed.), *The Social Impact of Computers* (3rd ed., San Diego: Elsevier, 2004).

Schneier, Bruce, *Data and Goliath: The Hidden Battles to Collect Your Data and Control Your World* (New York: W. W. Norton, 2015).

Searle, John R., "Minds, Brains and Programs," *Behavioral and Brain Sciences* 3, no. 3 (1980): 417–457.

Sennett, Richard, *Building and Dwelling: Ethics for the City* (London: Allen Lane, 2018).

Smolan, Sandy, "The Human Face of Big Data," PBS Documentary, February 24 2016, 56 mins, https://www.youtube.com/watch?v=kAZ8lK224Kw.

Stahl, Bernd Carsten, and David Wright, "Ethics and Privacy in AI and Big Data: Implementing Responsible Research and Innovation," *IEEE Security & Privacy* 16, no. 3 (2018): 26–33.

Susser, Daniel, Beate Roessler, and Helen Nissenbaum, "Technology, Autonomy, and Manipulation," *Internet Policy Review* 8, no. 2 (2019): 1–22.

Taipale, Kim A., "Data Mining and Domestic Security: Connecting the Dots to Make Sense of Data," *Columbia Science and Technology Law Review* 5, no. 2 (2003): 1–83.

TAPAC, "Safeguarding Privacy in the Fight Against Terrorism," Report of the Technology and Privacy Advisory Committee to the Department of Defense, March 1, 2004, https://purl.access.gpo.gov/GPO/LPS52114.

Thompson, Nicholas, and Ian Bremmer, "The AI Cold War that Threatens Us All," *Wired*, October 23, 2018, https://www.wired.com/story/ai-cold-war-china-could-doom-us-all/.

Wachter, Sandra, and Brent Daniel Mittelstadt, "A Right to Reasonable Inferences: Re-thinking Data Protection Law in the Age of Big Data and AI," *Columbia Business Law Review*, no 2 (2019): 494–620.

Woolley, Sam, and Phil Howard (eds.), *Computational Propaganda: Political Parties, Politicians, and Political Manipulation on Social Media* (Oxford: Oxford University Press, 2017).

Zuboff, Shoshana, *The Age of Surveillance Capitalism: The Fight for a Human Future at the New Frontier of Power* (New York: Public Affairs, 2019).

Zuckermann, Ethan, "Die Antwort auf Überwachung heisst Unterwachung," *Die Zeit*, October 7, 2013.

Index